CW00726624

Sustainable Building Adaptation

Series advisors

Carolyn Hayles, University of Bath
Richard Kirkham, University of Manchester
Andrew Knight, Nottingham Trent University
Stephen Pryke, University College London
Steve Rowlinson, University of Hong Kong
Derek Thomson, Loughborough University
Sara J. Wilkinson, University of Technology, Sydney

Innovation in the Built Environment (IBE) is a new book series for the construction industry published jointly by the Royal Institution of Chartered Surveyors and Wiley Blackwell. It addresses issues of current research and practitioner relevance and takes an international perspective, drawing from research applications and case studies worldwide.

- Presents the latest thinking on the processes that influence the design, construction and management of the built environment

- Based on strong theoretical concepts and draws on both established techniques for analysing the processes that shape the built environment – and on those from other disciplines

- Embraces a comparative approach, allowing best practice to be put forward

- Demonstrates the contribution that effective management of built environment processes can make

Books in the IBE series

Akintoye & Beck: *Policy, Finance and Management for Public-Private Partnerships*
Booth, Hammond, Lamond & Proverbs: *Solutions for Climate Change Challenges in the Built Environment*
Boussabaine: *Risk Pricing Strategies for Public-Private Partnership Projects*
Kirkham: *Whole Life-Cycle Costing*
London: *Construction Internationalisation*
Lu & Sexton: *Innovation in Small Professional Practices in the Built Environment*
Pryke: *Construction Supply Chain Management: Concepts and Case Studies*
Roper & Borello: *International Facility Management*
Senaratne & Sexton: *Managing Change in Construction Projects*
Wilkinson, Remøy & Langston: *Sustainable Building Adaptation*

For full details of the *Innovation in the Built Environment* series, please go to http://eu.wiley.com/WileyCDA/Section/id-811341.html

We welcome proposals for new, high-quality, research-based books which are academically rigorous and informed by the latest thinking; please contact Madeleine Metcalfe.

Madeleine Metcalfe
Senior Commissioning Editor
Wiley Blackwell
9600 Garsington Road
Oxford OX4 2DQ
mmetcalfe@wiley.com

Sustainable Building Adaptation: Innovations in Decision-Making

Sara J. Wilkinson

Associate Professor of Property and Construction
University of Technology
Sydney
Australia

Hilde Remøy

Assistant Professor of Real Estate Management
Faculty of Architecture
Delft University of Technology
The Netherlands

Craig Langston

Professor of Construction and Facilities Management
Bond University
Gold Coast
Australia

This edition first published 2014
© 2014 by John Wiley & Sons, Ltd

Registered Office
John Wiley & Sons, Ltd, The Atrium, Southern Gate, Chichester, West Sussex, PO19 8SQ, United Kingdom.

Editorial Offices
9600 Garsington Road, Oxford, OX4 2DQ, United Kingdom.
The Atrium, Southern Gate, Chichester, West Sussex, PO19 8SQ, United Kingdom.

For details of our global editorial offices, for customer services and for information about how to apply for permission to reuse the copyright material in this book please see our website at www.wiley.com/wiley-blackwell.

The right of the author to be identified as the author of this work has been asserted in accordance with the UK Copyright, Designs and Patents Act 1988.

All rights reserved. No part of this publication may be reproduced, stored in a retrieval system, or transmitted, in any form or by any means, electronic, mechanical, photocopying, recording or otherwise, except as permitted by the UK Copyright, Designs and Patents Act 1988, without the prior permission of the publisher.

Designations used by companies to distinguish their products are often claimed as trademarks. All brand names and product names used in this book are trade names, service marks, trademarks or registered trademarks of their respective owners. The publisher is not associated with any product or vendor mentioned in this book.

Limit of Liability/Disclaimer of Warranty: While the publisher and author(s) have used their best efforts in preparing this book, they make no representations or warranties with respect to the accuracy or completeness of the contents of this book and specifically disclaim any implied warranties of merchantability or fitness for a particular purpose. It is sold on the understanding that the publisher is not engaged in rendering professional services and neither the publisher nor the author shall be liable for damages arising herefrom. If professional advice or other expert assistance is required, the services of a competent professional should be sought.

Library of Congress Cataloging-in-Publication Data

Wilkinson, Sara, 1961–
  Sustainable building adaptation : innovations in decision-making / Sara J Wilkinson, Hilde Remøy, Craig Langston.
    pages cm
  Includes bibliographical references and index.
  ISBN 978-1-118-47710-6 (cloth)
1. Sustainable buildings–Design and reconstruction. 2. Buildings–Remodeling for other use.
3. Architecture–Conservation and restoration–Case studies. I. Remøy, Hilde Therese, 1972– II. Langston, Craig A. III. Title.
  TH880.W53 2014
  690.028′6–dc23

                                                          2013030498

A catalogue record for this book is available from the British Library.

Wiley also publishes its books in a variety of electronic formats. Some content that appears in print may not be available in electronic books.

Cover design by Andy Meaden, Meaden Creative

Set in 10/12pt Sabon by SPi Publisher Services, Pondicherry, India
Printed and bound in Malaysia by Vivar Printing Sdn Bhd

1 2014

Contents

**Part III   Adaptation Decision-Making
             and Optimisation   183**
Craig Langston

About the Authors

Dr Sara J. Wilkinson is Associate Professor of Property and Construction at the University of Technology Sydney, Faculty of Design Architecture and Building, Sydney, Australia. She has a combination of professional industry and academic experience spanning more than 30 years. The research described in Part I: Building Adaptation is the result of work undertaken over a 16-year period and has been funded by Jones Lang LaSalle and the Royal Institution of Chartered Surveyors. Sara's research focus is building adaptation within the context of sustainability and represents areas of professional practice prior to becoming an academic. Her PhD examined building adaptation and the relationship to property attributes, whilst her MPhil explored the conceptual understanding of green buildings. Sara is a member of the RICS Oceania Sustainability Working Group. She is the International Federation of Surveyors (FIG) Vice-Chair of Commission 10 'Construction Management & Construction Economics'. Sara is also the author of eight books/book chapters and was awarded the RICS COBRA Conference Best Paper Award in 2012 for her paper 'The increasing importance of environmental attributes in commercial building retrofits', RICS COBRA, Las Vegas, NV, USA. September 2012. http://www.rics.org/au/knowledge/research/conference-papers/cobra-2012-environmental-attributes-in-commercial-building-retrofits/

Dr Hilde Remøy is Assistant Professor of Real Estate Management at the Faculty of Architecture, Delft University of Technology, Delft, the Netherlands. She has experience with adaptive reuse from both practice and academia. The research described in Part II: Adaptive Reuse is the result of studies undertaken in the period 2005–2013. Hilde's research focus is adaptive reuse of existing buildings that have lost their original function, related to obsolescence and vacancy of existing buildings and locations. In research and education, she works on studies concerning the influence of physical property characteristics on obsolescence and adaptive reuse potential and studies to define the future value of reused buildings and cultural heritage. Hilde is the author of several books/book chapters.

Dr Craig Langston is Professor of Construction and Facilities Management at Bond University's School of Sustainable Development, Gold Coast, Australia. He has a combination of industry and academic experience spanning more than 35 years. The research described in Part III: Decision-making and

Optimisation is the result of three Australian Research Council (ARC) Linkage Project grants comprising:

- 2009–2012 – Langston, C., Smith, J., Herath, G., Datta, S., Doloi, H. and Crawford, R.H., Making Better Decisions about Built Assets: Learning by Doing, ARC Linkage Project $180,000 LP0990261 (Industry partners: Williams Boag Architects and Assetic Australia).
- 2007–2010 – Langston, C., Liu, C., Beynon, D. and de Jong, U., Strategic Assessment of Building Design Adaptive Reuse Opportunities, ARC Linkage Project $210,000 LP0776579 (Industry partners: Williams Boag Architects and The Uniting Church in Australia).
- 2006–2009 – Crawford, R.H., Datta, S. and Langston, C., Modelling Environmental and Financial Performance of Construction. Sustainability Innovation Feasibility Tool. ARC Linkage Project $179,000 LP0667653 (Industry partner: Williams Boag Architects).

Craig is the author of five international books. In 2010, he won the Bond University Vice-Chancellor's Quality Award for Research Excellence. He was awarded the Emerald Literati Network Award for Excellence in 2013 for his paper 'Validation of the adaptive reuse potential (ARP) model using iconCUR', Facilities, 30(3–4), 105–123 (2012).

Preface: The Rise of Building Adaptation

A point has been reached in history when we must shape our actions throughout the world with a more prudent care for their environmental consequences. Through ignorance or indifference we can do massive and irreversible harm to the earthly environment on which our life and well-being depend. Conversely, through fuller knowledge and wiser action, we can achieve for ourselves and our posterity a better life in an environment more in keeping with human needs and hopes... To defend and improve the human environment for present and future generations has become an imperative goal for mankind.[1]

It is four decades since these words of forewarning were written, and we should reflect on whether we have heeded the statements. The declaration is a warning and a call to action. The proclamation asserts that humans need to be more prudent in respect of the environment; yet in those 40 years, greenhouse gas emissions have increased, pollution has worsened, and social inequity and injustice around the world has continued to attract global attention. If anything, the environmental legacy for future generations is less than it was in 1972.

The challenge of achieving sustainable development in the twenty-first century will be won or lost in the world's urban centres, and this is due to the contribution that the built environment makes to greenhouse gas emissions and global warming. The challenge is immense and overwhelming, both in terms of its magnitude and potential consequences if humankind does not adapt its behaviours towards the environment. Climate change impacts are occurring, disproportionately affecting developing nations, and are projected to get much worse over time. It is expected that there will be increased variability in climate events, such as harder and more frequent storms, which will lead to changes in climatic averages such as increased water scarcity. Globally as humankind adapts and evolves its behaviours and government strategies and policies, we are transitioning from the 'industrial age' to the 'ecological age'.

The built environment, if upstream emissions from heat and electricity are included, is responsible for around 45% of total global greenhouse gas emissions (GGE). Also there are impacts from water and resources consumption within buildings. As commercial buildings have a life cycle measured in decades or even centuries, the existing stock is of particular interest and consequence. Significantly, our window of opportunity for pre-emptive action to avoid higher levels of climate change and temperature increase is to act decisively up to 2050; time is not on our side. When compared to other sectors, such as transport or waste, the contribution of sustainable building adaptation to climate change mitigation is abundantly clear.

With 1–2% of new buildings added to the total stock annually, humankind needs to adapt its existing buildings, and quickly. While all new construction should adopt sustainability features in design and operation, given typical rates of replacement much of the built environment that will exist in 2050 has already been built. Furthermore, the Inter Governmental Panel on Climate Change (IPCC) concluded that:

> Over the whole building stock the largest portion of carbon savings by 2030 is in retrofitting existing buildings and replacing energy using equipment due to the slow turnover of the stock.[2]

The greatest challenge is the development of successful strategies for adapting existing buildings due to their slow turnover; in other words, effective decision-making for sustainable building adaptation is critical to deliver needed building-related GGE reductions globally. Many cities have acknowledged this need to act and have developed and adopted strategies aimed to deliver carbon neutrality within fixed periods. Local government authorities are encouraging sustainable building adaptation to lower building-related energy consumption and associated emissions.

Sustainable adaptation of existing stock is a universal concern that increasing numbers of local, state and national governments must endeavour to address within the short to medium term. In most developed countries, more is now spent on building adaptation (including maintenance, repair, retrofit and reuse) than new construction, and this represents a gradual but consistent change from decades of investment dominance in new-build projects. There is a need for greater knowledge and awareness of what happens to society's buildings over time and how we might adapt them sustainably. This action includes avoiding premature destruction through finding new uses for buildings that have become unwanted or obsolete. While new development must also be sustainable, there is insufficient time for us to act unless proactive intervention into the performance of existing building stock becomes a priority.

This research-based book contributes significantly to a more informed understanding and management of decisions relating to the sustainable adaptation of existing commercial buildings. This work collectively offers guidance towards a balanced approach that incorporates sustainable and optimal approaches for effective management of sustainable adaptation of existing commercial buildings. It is divided into three discrete parts concerning building adaptation, adaptive reuse, and adaptation decision-making and optimisation.

Part I has been written by Dr Sara J. Wilkinson. She establishes the definition of adaptation in the context of this book. She reviews and synthesises the key literature, while progressively developing the research questions, hypotheses and a conceptual model towards a knowledge-based approach to sustainable office adaptation. She describes and substantiates her latest research demonstrating how to make a preliminary assessment of adaptation potential using the Melbourne CBD as an illustrative case study. A large focus for this part concerns the connection between sustainability and building adaptation.

Part II has been written by Dr Hilde Remøy. She presents her research conducted into Dutch office change of use adaptations. Adaptive reuse, defined as significant functional change applied to obsolete buildings as an alternative to premature destruction, is her focus. Many exemplars demonstrating application of this approach in the Netherlands are provided and augmented with a number of international case studies. In this part, the relationship of adaptation, retrofitting, alteration and inherent flexibility provided by the initial design solution is explored, including discussion of the practical lessons learned from the underpinning work (as case studies for the practitioner audience) and a clear statement of the theoretical contributions involved.

Part III has been written by Dr Craig Langston. He covers adaptation decision-making and optimisation using multiple criteria. He describes and substantiates his research into how to make a strategic assessment of whether and when to adapt. Cost planning is a key feature of the decision-making process and its integration into a broader financial–social–environmental frame is explored. He also introduces a model to assess new design to ensure that it will deliver adaptation benefits much later in life. Each presented decision/optimisation model is demonstrated via one or more actual case studies.

To sum up, the key issue and motivation for this book is that we need to adapt our existing building stock to reduce its environmental footprint, to aim for higher sustainability, better energy performance and more efficient use of natural resources. We are currently some way from this being standard practice in many urban settlements. Whilst there are an abundance of environmental rating tools to choose from across a range of countries, there is patchy take-up within the real estate markets, especially with lower quality or lower profile stock. Nevertheless, there is an increasing amount of legislation relating to sustainability and evidence that industry practices are improving – but whether the rate of uptake is sufficient to make a meaningful change only time will tell.

As is often quoted, 'the greenest buildings are the ones we already have'.[3]

Notes

1   Extract from the Declaration of the UN Conference on the Human Environment (1972), available online at http://www.unep.org/Documents.Multilingual/Default.Print.asp?documentid=97&articleid=1503&l=fr. Accessed 19 August 2013.

2   Extract from the Intergovernment Panel on Climate Change (IPCC) Fourth Assessment Report (2007), available online at http://www.ipcc.ch/publications_and_data/ar4/wg3/en/ch6-ens6-es.html. Accessed 19 August 2013.

3   Originally attributed to Jacobs, J. (1961) *The death and life of great American cities*, New York: Random House.

Part I Building Adaptation

The author for this part is Dr Sara J. Wilkinson. Sara is Associate Professor of Property and Construction at the University of Technology Sydney, Faculty of Design Architecture and Building, Sydney, Australia. She has a combination of professional industry and academic experience spanning more than 30 years.

The research described in this part is the result of work undertaken over a 16-year period and has been funded by Jones Lang LaSalle and the Royal Institution of Chartered Surveyors. Sara's research focus is building adaptation within the context of sustainability, and represents areas of professional practice prior to becoming an academic. Her PhD examined building adaptation and the relationship to property attributes, whilst her MPhil explored the conceptual understanding of green buildings. Sara is a member of the RICS Oceania Sustainability Working Group. She is the International Federation of Surveyors (FIG) Vice-Chair of Commission 10 'Construction Management & Construction Economics'. Sara is also the author of eight books/book chapters and was awarded the RICS COBRA Conference Best Paper Award in 2012 for her paper 'The increasing importance of environmental attributes in commercial building retrofits', RICS COBRA, Las Vegas, NV, USA. September 2012. http://www.rics.org/au/knowledge/research/conference-papers/cobra-2012-environmental-attributes-in-commercial-building-retrofits/

This part of the book establishes the definition of adaptation within the context of this book. It reviews and synthesises the relevant literature, while progressively developing the research questions, hypotheses and the conceptual model towards a knowledge-based approach to sustainable office adaptation.

Sustainable Building Adaptation: Innovations in Decision-Making, First Edition.
Sara J. Wilkinson, Hilde Remøy and Craig Langston.
© 2014 John Wiley & Sons, Ltd. Published 2014 by John Wiley & Sons, Ltd.

It describes and substantiates latest research demonstrating how to make a preliminary assessment of adaptation potential using Melbourne as an illustrative case study. Further, this part covers the issue of decision-making in commercial building adaptation. It uses empirical data to identify and explore the factors that are most important in adaptation and how they relate to sustainability. Whereas many previous studies relied on relatively small data sets of adaptation on which to base models and findings, this research is built on a significant number of cases over an extended time period.

Chapter 1 commences with a definition of building adaptation and alternate terms. Sustainability is explored within the context of social, economic and environmental paradigms. The relationship between building life cycles and adaptation is also explained and how it can affect the timing and degree of adaptation. The various decision options and different levels of adaptation are illustrated to demonstrate the numerous options available.

Chapter 2 describes the drivers and barriers for adaptation. Building life cycle theory is introduced and the ways in which adaptation occurs at different stages after completion. These adaptations may occur as a result of legal, economic, physical, social and environmental drivers. The relevance of building performance theory to adaptation is explained in this chapter as well as how performance inevitably declines over time. In the context of the social, environmental and economic factors, the links between building adaptation and sustainability are then highlighted. Finally, other attributes associated with adaptation, such as physical, locational, land use and legal attributes, are discussed.

Chapter 3 focuses on how to assess adaptation using a robust method developed to identify the most important attributes associated with adaptation. Using a large database of adaptation events, principal component analysis is undertaken to establish which attributes are most important. From this analysis a Preliminary Adaptation Assessment Model (PAAM) is developed. Critically this model is designed for non-experts to use in making an initial assessment of a building's potential for minor adaptation. The chapter concludes with an illustrative case study to demonstrate the application of the model in practice.

Chapter 4 uses case studies to explore sustainable building adaptation in Melbourne, Australia. The City of Melbourne is committed to encouraging sustainable adaptation through its innovative 1200 Buildings Program. This chapter identifies the measures typically adopted in sustainable building adaptation before describing ten sustainable building adaptations. The case studies highlight the rationale and objectives for each adaptation, their sustainable features, key challenges and the outcomes of adaptation. The remainder of the chapter compares the adaptations with regards to a number of attributes previously shown to be important.

Defining Adaptation

1.1 Introduction

This chapter defines adaptation and alternate terms commonly adopted around the world. The distinctions between in-use and across-use adaptations are identified before describing the significance of adaptation within the context of sustainability. Sustainability is discussed to illustrate why the need to adapt our existing stock becomes more of an imperative as time passes. Adopting the standard convention, sustainability is explored with the context of environmental, social and economic paradigms. The chapter then moves on to show the relationship between adaptation and building life cycles and how this can vary the timing and extent of adaptation projects.

Contextual placing of adaptation within our systems of governance is then discussed with reference to the drive for climate change adaptation such as carbon neutrality that is prompting city authorities around the world to implement legislation and policy to encourage sustainable building adaptation. The scope and extent of these initiatives will increase as the manifestations of anthropogenic climate change become more apparent with the passage of time. The framework currently adopted in Melbourne, Australia, is used to illustrate what is being done in this respect.

The final section examines the stakeholders and decision-making issues in sustainable building adaptation and how they affect the degree of adaptation and sustainability that may be achieved. The numerous levels of adaptation as well as the different stakeholders can make the possible outcomes vary extensively.

Sustainable Building Adaptation: Innovations in Decision-Making, First Edition.
Sara J. Wilkinson, Hilde Remøy and Craig Langston.
© 2014 John Wiley & Sons, Ltd. Published 2014 by John Wiley & Sons, Ltd.

1.2   Terminology

Adaptation, in the context of buildings, is a term that has been broadly interpreted and defined by many researchers (Ball 2002; Mansfield 2002; Douglas 2006; Bullen 2007). Adaptation is derived from the Latin 'ad' (to) 'aptare' (fit). Typically the definitions refer to 'change of use', maximum 'retention' of the original structure and fabric of a building as well as extending the 'useful life' of a property (Ball 2002; Mansfield 2002; Douglas 2006; Bullen 2007). Frequently there are terms such as renovation, adaptive reuse, refurbishment, remodelling, reinstatement, retrofitting, conversion, transformation, rehabilitation, modernisation, re-lifing, restoration and recycling of buildings used to define adaptation activities. The terms all have different meanings, for example, 'refurbishment' comes from the word refurbish which means, 're', to do again and, 'furbish', to polish or rub up. On the other hand, 'conversion' literally means to convert or change from one use to another, for example, a barn converted to a residential property, and this aspect of adaptation is dealt with specifically in Part II. Three decades ago Markus (1979) noted these terms existed in an 'unhappy confusion'; it is an unhappy confusion which still exists and one we must be cognisant of.

Building adaptation occurs 'within use' and 'across use'; that is, an office can undergo adaptation and still be used as an office (i.e. within-use adaptation), or it may change use to residential ('across-use' adaptation) (Ellison and Sayce 2007). Adaptation of existing buildings can encompass some or all of the terms renovation, adaptive reuse, refurbishment, remodelling, reinstatement, retrofitting, conversion, transformation, rehabilitation, modernisation, re-lifing, restoration and recycling of buildings. For this part of the book, a broad definition is adopted, which includes all forms of adaptation, except for minor day-to-day repair and maintenance works. A useful definition of building adaptation, adopted for this book, is

> any work to a building over and above maintenance to change its capacity, function or performance, in other words, 'any intervention to adjust, reuse, or upgrade a building to suit new conditions or requirements'. (Douglas 2006:4)

1.3   The Significance of Building Adaptation

With the rise in consensus within the scientific community regarding anthropogenic activity and climate change, increased sustainability in the built environment is an imperative (Stern 2006; Garnaut 2008). One method of reducing mankind's environmental impact is to adapt buildings rather than default to demolish and new build. This book examines the case for adaptation, adaptive reuse and decision-making with regard to the building adaptation.

Buildings are inextricably linked to sustainability issues, and the construction industry has a major role in reducing the adverse effects on the environment as buildings contribute around half of all greenhouse gas emissions (UNEP 2006). Sustainability has a broad and differing definition depending upon the context in which it is used. It is most commonly defined as 'meeting the needs of the present without compromising the ability of future generations to meet their own needs' (WCED 1987:2) or 'using, conserving and enhancing the community's resources so that ecological processes, on which life depends, are maintained, and the total quality of life, now and in the future, can be increased' (Commonwealth of Australia 1992). Brundtland (WCED 1987) described the concept of sustainable development as a strategy to optimise the relationship between the global society and its natural environment with consideration of the social, economic and environmental goals of society.

International concern for the environment was reflected via the UN conference in Stockholm in 1972 and the idea of eco-development emerged as 'an approach to development aimed at harmonizing social and economic objectives with ecologically sound management' (Gardner 1989). Although eco-development was the precursor of the concept of sustainability, the early concept of sustainable development was firmly entrenched within the environmental movement, and sustainability was often interpreted as sustainable use of natural resources (Hill and Bowen 1997). Debate continued on the appropriate definition of the concept of sustainability. It was argued that development inevitably leads to some drawdown of stocks of non-renewable resources and that sustainability should mean more than the preservation of natural resources (Solow 1993), while it was believed that sustainability had three dimensions, those of environmental, social and economic sustainability (Goodland 1995; Elkington 1997). The divergence of opinions demonstrated that sustainability is so broad an idea that a single definition cannot capture the concept; however, there is agreement that uncontrolled exploitation of natural resources is not beneficial to humankind in the long term (Hill and Bowen 1997). It was proposed that sustainable construction meant 'creating a healthy built environment using resource-efficient, ecologically-based principles' (Kibert 2005). Four principles, adopted by Hill and Bowen (1997) in the concept of sustainable construction, were social sustainability, economic sustainability, technical sustainability and biophysical sustainability. This notion of sustainable construction provides the building and construction industry with a practical framework to guide the implementation of sustainable buildings (Hill and Bowen 1997).

Adaptation is inherently environmentally sustainable because it involves less material use (i.e. resource consumption), less transport energy, less energy consumption and less pollution during construction (Johnstone 1995; Bullen 2007). The embodied energy within existing stock is considerable, and the Australian Greenhouse Office (AGO) estimated the reuse of building materials saves approximately 95% of embodied energy (Binder 2003). Even when economic costs for adaptation are high, the environmental argument along with social factors may sway the decision in favour of adaptation (Ball 2002). The process of demolition is a wasteful process in terms of materials unless they are reused or recycled (Department of the Environment and Heritage 2004). Since the late 1990s the concept of sustainability has been one of the major drivers of adaptation due to the notion of recycling of buildings (Ball 2002). Upgrading performance of existing stock, through adaptation, is the most critical aspect of improving sustainability of the built environment (Cooper 2001).

Humans have adapted buildings since they started constructing. Over time, the usefulness of any building for its original function diminishes; this process is known as obsolescence and represents a lack of utility. Obsolescence takes several forms such as physical obsolescence, where buildings or their component parts literally wear out. Functional obsolescence occurs where the original function of the building becomes redundant, for example, the workhouses built in the Victorian period throughout England for the poor and destitute are no longer perceived as appropriate methods of housing people experiencing economic hardship and unemployment. Economic obsolescence occurs when the economic rationale for a building is removed; an example is the 2007 closure in Geelong, Victoria, of the Ford Motor Company factory as a result of cheaper production elsewhere and a downturn in vehicle sales generally. Locational obsolescence occurs when the location of the building is no longer suitable, such as warehouses sited on canals in England that became obsolete when motorways overtook canals as rail and road primary means of transporting goods and materials in the nineteenth century.

Obsolescence can affect any building at any time during its life cycle and can trigger an opportunity for adaptation. Building obsolescence is the subject of much research (Cowan et al. 1970a, b, c; Nutt et al. 1976; Baum 1991; Building Research Board 1993; Khalid 1994; EKOS Limited and Ryden Property Consultants 2001). Previous studies examined the causes and impact of building obsolescence and ways to defer the time when a building has no utility whatsoever and demolition remains the only viable option. One way of deferring obsolescence in buildings is to adapt them either through a change of use or within the existing use (Kincaid 2002). Selected examples from Hong Kong (China) are provided in Figure 1.1, Figure 1.2 and Figure 1.3. Further discussion of obsolescence is contained in Part II.

Substantial expenditure is directed to building adaptation across developed nations, and in the UK more work is undertaken on adaptation than new build (Egbu 1997; Ball 2002). Half of the total expenditure on construction in the UK was on existing buildings (Cooper 2001), and in 2004 £45 billion was spent on UK building adaptation (Goodier and Gibb 2004). Looking at the Australian built environment, construction normally contributes between 5% and 6% of national Gross Domestic Product (i.e. 6.7% in 2002/2003) (Australian Bureau of Statistics 2013). A median percentage of 17.8% of all construction work undertaken in Australia for the decade between 1991 and 2001 was on existing buildings. With an estimated $267 billion of new commercial property to be built in Australia before 2018, the performance gap between new and old stock looks set to increase (Romain 2008). The proportion and amount of annual expenditure on building adaptation in Australia and other national economies of developed countries demonstrates the importance of adaptation to business and commerce, both in the past and increasingly into the future. Similar circumstances for stock

Figure 1.1   Western Market (1906–2003), Sheung Wan, Hong Kong.

Figure 1.2   60–66 Johnston Road (1888–2008), Wan Chai, Hong Kong.

Figure 1.3 Former Marine Police Headquarters (1881–2009), Tsim Sha Tsui, Hong Kong.

condition and proportions of expenditure on adaptation exist in other developed nations globally.

'Highest and best use' is defined as 'the use which results in the most efficient and/or profitable use' of the building (API 2007). Highest and best use is a key appraisal and zoning principle employed in valuing land or buildings and is an important influencing factor in determining obsolescence. Clearly the value of a building and its use are linked closely: highest and best use leading directly to highest present value providing the greatest return for investors and owners. Furthermore, a building's value is influenced by the surrounding environment, so land use has to be consistent or complimentary to neighbouring land uses. Other influencing factors affecting value are local competition and political forces; therefore, the timing of a development or redevelopment is vital to achieve the highest and best use. In city centres, multiple uses are not uncommon, and this makes an appraisal of highest and best use more complex given a combination of land uses may be optimal. Highest and best use appraisals consider four factors: legal permissibility, physical possibility, financial feasibility and the maximum productive use (API 2007). Even so the most comprehensive appraisal is only relevant to a specific point in time and is an expert opinion only (Tosh and Rayburn 2002). Some argue for consideration of social criteria and not only economic factors in the appraisal (Jarchow 1991; Nahkies 2002). Over time the methods of accounting for the costs and benefits of social and community aspects in development have been acquired and may be integrated into the assessment. In this book the underlying assumption is that adaptation is predicated on the goal of achieving and maintaining highest and best use for a building at a given point in time.

Globally, the market is noting an increasing amount of adaptation in buildings over the last 20 years. For example, increased levels of adaptation were noted in Alabama and Chicago (Olson 2005; Colchimario 2006). There was a 'frantic pace' of adaptation activity in the UK that outstripped new building activity (Kincaid 2000). A study into the adaptation of offices to residential uses found a large upturn in activity in Boston, Sydney, Melbourne and Vancouver during the 1990s (Heath 2001), while a UK study reported an increased level of adaptation in the retail sector (Douglas 1994). This increase in the rate and amount of adaptation across developed countries is, in part, a response to the case that adaptations are typically faster to complete and occupy than new build and that adaptation often costs less (Chandler 1991; Highfield 2000). Subsequent to the 2008 global financial crisis, there has been a slowdown in all areas of construction. Since the early 2000s there has been a discernible response to the emerging importance of sustainability within the built environment and embodied energy within existing buildings, and thus adaptation can represent a more sustainable solution to new build (Bullen 2007). This move towards incorporating sustainability has occurred alongside significant UK and Australian Government-led global economic reports (Stern 2006; Garnaut 2008) highlighting the potential outcomes of ignoring global warming and climate change. The momentum for sustainability in buildings has been further increased through the adoption of corporate social responsibility (CSR) reporting by leading business organisations around the world and the subsequent adaptations that enhance the sustainability of their building stock (Newell and Sieraki 2009).

The age and quality of the stock in an area affect the amount and scope of adaptation undertaken. Within Australia, previous studies estimate between 85 and 90% of the commercial building stock is aged over 10 years (Davis Langdon 2008). In older more established cities, the average age of the building stock is higher still. Furthermore, in established urban centres, only small percentages of new buildings are added to the existing stock total each year. For example, in London 1–2% is added to the total stock of commercial buildings annually (Knott 2007), whereas Melbourne typically has 2–3% added to total stock (Jones Lang LaSalle 2008). It is estimated that 87% of the residential stock the UK will have in 2050 is already built and 89% of the stock Sydney will have in 2030 is already built (Kelly 2008). As a result there is an ongoing need to adapt the existing stock to meet the changing current and future needs of investors and building users. Pressure is placed on existing building stock in Australia from increased immigration, resulting in further opportunity to adapt buildings (Foran and Poldy 2002). Melbourne is seeking to grow to a population of five million by the year 2025, and recent growth exceeded this prediction (Department of Premier and Cabinet 2008). The UK also experienced population growth in the first decade of the twenty-first century through immigration within an expanding European Union. Population increases put pressure on the existing stock to meet societal needs that can make adaptation attractive. Moreover, the situation is compounded with a construction labour skill shortage in Australia which has driven new build construction costs upwards, and adaptation can be an attractive economic alternative in some cases.

Globally in some urban areas, vacancy rates for office buildings are high and rates are increasing with the ongoing global economic turmoil (RICS 2009). Vacancy rates are higher for lower-grade stock, and some sections of stock that have been vacant for three or more years are considered to be a structural and long-term issue (Remøy and Van der Voordt 2006). For instance, in the Netherlands there was seven million square metres of office space vacant in 2012, which presented challenges to owners by way of lost income (DTZ 2012). Moreover, empty buildings are more vulnerable to vandalism, arson and squatting that drive up ownership costs and, in the long term, the effects of vacancy including social blight and economic decline. It was concluded that one million square metres of Dutch office space (or 2% of total stock) should be removed from the market because it is outdated and suffering to various degrees from technical, functional, locational and physical obsolescence (Remøy and Van der Voordt 2006). In Australian cities the positive economic conditions of the early 2000s have led to comparatively low vacancy rates. Melbourne had an all-time low vacancy rate for office space in 2007 when rates were 4.7% (Savills 2009). However, as a result of the global financial crisis in 2008, national office vacancy rates increased across Australian cities to 9% in July 2011 (PCA 2013). Adaptation offers a new economic life for a building at a fraction of the cost of new construction, and with a greater amount of lower-grade space available, there is an opportunity for businesses to occupy better-quality space as developed countries move out of recession in due course.

1.4   Decision-Making Issues in Building Adaptation

Building adaptation decision-making is complex (Blakstad 2001; Douglas 2006). There are many stakeholders involved, each representing a different perspective. Decision-makers are investors, producers, developers, regulators, occupants/users and marketers (Kincaid 2002). An additional layer of complexity is that these stakeholders make decisions at different stages in the process and each has different degrees of influence (see Table 1.1).

Generally decisions made at the early stages of the process have an ongoing impact throughout the project. For example, the decision to change the use affects all the decisions that follow. Furthermore, the capacity of stakeholders to influence decisions can be classed as either direct or indirect. Another layer is added where a stakeholder intends to be an occupier or user, in which case the decisions will have a daily impact on their ongoing business operations. The motivations of stakeholders influencing decision-making vary, for example, a developer who intends on selling the property post-adaptation experiences different drivers than if the intention is to retain the property within the developer's property portfolio. In summary, stakeholders are multiple and exert their influence to different degrees at different stages.

Table 1.1 Decision-makers in building adaptation.

Decision-makers	Professional and other affiliations	Stage in adaptation where decisions are made
Investors	Pension/superannuation funds, insurance companies, banks, independent investors, professionals who find capital to invest	Beginning/early
Producers	Professional team – facilities manager, quantity surveyor, architects, engineers, contractors, surveyors, suppliers, fire engineers, structural and mechanical and electrical engineers	Quantity surveyor/architect at feasibility stage Design stage Construction stage
Marketeers	Surveyors, stakeholders, professionals who find users for buildings	During design (if selling off plan) and/or construction stage
Regulators	Local authorities, planners, heritage, building surveyors, fire engineers	During design stage (and possibly during construction if amendments are made)
Policymakers	Federal, state and local government departments	Indirect effect on decision-making in adaptation at all stages
Developers	Organisations that combine investment, production and marketing in whole or in part. Professionals from aforementioned bodies and others	Beginning/early
Users: corporate, residential	Large institutional owners and users, individuals, business organisations and occupiers	—

1.5   Decision Options and Levels of Adaptation

A further aspect is the range of options available to stakeholders (Kincaid 2002). Kincaid rationalised the options as follows: Option one is to change the use with minimum intervention because of the inherent 'flexibility' of the building. Option two is for adaptation with minor change, while option three requires a higher degree of intervention and is typically referred to as 'refurbishment' or 'retrofitting'. Option four involves selected demolition, whereas option five is the extension of the facility. Finally, option six is demolition and redevelopment and is selected when the social, economic, environmental, regulatory and physical conditions are such that the building is at the end of the life cycle, lacking utility (Bottom 1999). This part of the book is focused on decision-making that occurs through options two to five inclusive. Effective decision-making demands the consideration of issues such as framing the issue properly, identifying and evaluating the alternatives and selecting the best option (Turban et al. 2005; Luecke 2006).

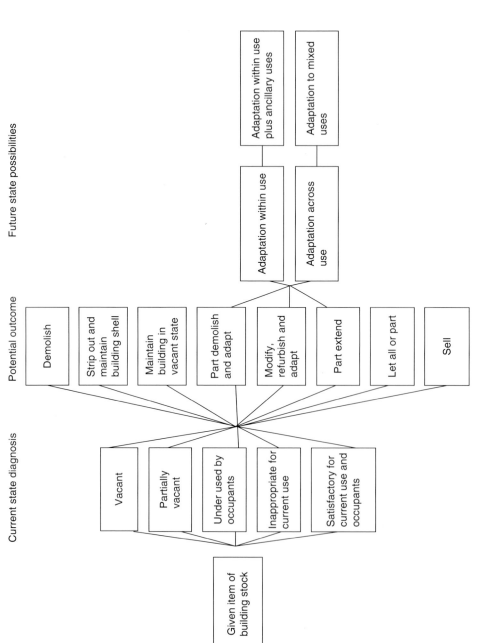

Figure 1.4 Options for adaptation.

Others have noted different option outcomes for adaptation such as rebuild, demolish and refurbish (Ohemeng 1996; Arup 2008). The 'options for adaptation' model is shown in Figure 1.4 and represents identification of all the potential outcomes found. Any building, for example, could be satisfactory for its current use and occupants, inappropriate for current use, underused by occupants, partially vacant or vacant. Following on from any of these states, the potential outcomes for the building are as follows: demolish; strip out and maintain building shell; maintain building in vacant state; part demolish and adapt; modify, refurbish and adapt; part extend; let all or in part; or sell. Where the decision is to either part demolish and adapt; modify, refurbish and adapt; or part extend, choices need to be made between an adaptation within use and an across-use adaptation. Finally, an adaptation within use may add ancillary uses, and an across-use adaptation could include mixed uses (Kincaid 2002).

Another way of distinguishing different levels of building adaptation modifies an approach developed by Arup (2008) where level 1 is very minor adaptation, level 2 is 'alterations' adaptation such as fit-outs to individual floors, level 3 a 'change of use' adaptation and level 4 major alterations and possibly extensions known as 'alterations and extensions'. Adaptation works are more or less progressively more extensive with each level.

1.6   Adaptation and Different Land Uses

Land use is the anthropological usage of land that involves the management and modification of natural environment into built environment. Land use practices differ around the globe. It is a term used by urban planners, who undertake land use planning and regulate the use of land in an attempt to avoid land use conflicts. Land use plans are implemented through land division and use ordinances and regulations, such as zoning regulations.

As time passes, some land uses become obsolete. For example, industrial land uses change as technological innovation renders premises outdated. At this point there is an opportunity through changes to planning zones, to adapt the buildings through change of use. Without change of zoning, the options are limited to within-use adaptation.

Although land use practices vary from country to country, in developed nations broadly the framework is similar. There are regulations that have to be followed in order to change land use. Increasingly planning authorities are adopting sustainability as a goal, and their policies and strategies reflect these aspirations. Furthermore, many planning authorities are also responsible for transportation policy and endeavour to link urban development to transport planning. The following list gives an indicative overview of the terms adopted in some jurisdictions:

- *Land use*: the way that land is used by people
- *Zone(s)*: areas that allow certain land uses
- *Commercial zone*: businesses that sell goods and services to local citizens (retail) or other businesses (wholesale)

Table 1.2   Land uses in the Melbourne CBD.

1. Office, premium	18. Entertainment/recreation indoor
2. Office, A grade	19. Cultural and community use
3. Office, B grade	20. Conferences/meetings
4. Office, C grade	21. Flats/apartment/unit
5. Office, D grade	22. Hotel/motel
6. Ungraded office	23. Hostel/backpackers accommodation
7. Retail, premium	24. Private hotel/boarding house
8. Retail	25. Institutional accommodation
9. Retail – stand-alone shop	26. Serviced apartment
10. Wholesale	27. Student accommodation
11. Manufacturing	28. Corporate supplied accommodation
12. Workshop/studio	29. Student apartment
13. Equipment installation/plant room	30. Parking – private, covered
14. Transport	31. Parking – commercial, covered
15. Storage	32. Common area
16. Education/research	33. Showroom
17. Hospital/clinic	Gallery/museum/public display area

- *Industrial zone*: Factories, warehouses or manufacturing plants that produce mass quantities of a product
- *Residential zone*: places for people to live
- *Public/semipublic (civic) zone*: an area that serves or is used by all people
- *Parks/recreation zone*: an area of land reserved for public use and recreation
- *Agricultural zone*: land used for livestock, growing crops and farm buildings

Using Melbourne as an example, it is the state capital of Victoria, Australia, with approximately 3.6 million people or 72% of the state's population residing there (Australian Bureau of Statistics 2006). In June 2007 the population for the City of Melbourne was estimated to be 81,144 – a growth of 60% since 2001. Recent planning policies have increased residential accommodation in the city encouraging more people to live in the city; this is in common with other Australian and international cities. State and city government is located in the city, along with academic institutions, theatres, restaurants, arts centres, commercial head offices and business organisations, retail and recreational facilities. Table 1.2 shows 34 land uses in the Melbourne CBD. It is clear that many land uses exist, though sometimes similar, for example, office land use is split into six categories and retail has three classifications. Compared to the more generic list in Section 1.6, it is a good example of how different authorities classify land uses.

1.7   Conclusion

The definitions of adaptation and the concepts presented, analysed and discussed in this chapter are concerned with the decisions of separate stakeholders and their collective behaviour that is reflected in the form of our cities and towns globally. Clearly it is difficult to determine the reasoning behind the subjective decisions executed by individuals and collectively as a society, and this section of the book describes the environment in which adaptation decisions are effected. Other parts of the text examine decisions made in respect of different stakeholder groups or specific types of adaptation such as conversion or across-use adaptation.

The various terms used with regard to adaptation to a large extent reflect changes in fashion and terminology that inevitably occur over time and in different locations. For example, the term adaptation is a preferred US term, whereas the UK favours the word refurbishment. Wherever one comes across these terms, it is important to seek clarity as to exactly what work is planned to avoid confusion. This chapter has also shown that policymakers and legislators are increasingly developing and implementing strategies to mitigate the perceived impacts of climate change. Due to the contribution of and impact of the built environment on carbon emissions, adaptation is perceived to be an area where significant positive outcomes are possible.

Finally, the complexity of decision-making was described, a situation with multiple potential outcomes influenced by a number of stakeholders and decision-makers who exercise influence and power at different stages of the process. It is apparent that discussions and decisions in respect of adaptation and sustainability will undergo much iteration before final outcomes are realised. Clearly framework and decision-making tools are needed to assist stakeholders to make informed and, above all, effective decisions in respect of sustainable building adaptation for all our sakes.

References

API. (2007) *Valuation of real estate* (12th edition), Sydney: Australian Property Institute.

Arup(2008) *Existing buildings: survival strategies*, Melbourne: Arup.

Australian Bureau of Statistics (2006). 2006 QuickStats: Melbourne. Available at http://www.censusdata.abs.gov.au/ABSNavigation/prenav/ProductSelect?newproducttype=QuickStats&btnSelectProduct=View+QuickStats+%3E&collection=Census&period=2006&areacode=205&geography=&method=&productlabel=&producttype=&topic=&navmapdisplayed=true&javascript=true&breadcrumb=LP&topholder=0&leftholder=0&currentaction=201&action=401&textversion=false. Accessed on 1 May 2011.

Australian Bureau of Statistics. (2013). Available at http://www.abs.gov.au/ausstats/abs@.nsf/products/B50895CF382A55D5CA256E6700062A55 Accessed on 19th November 2013.

Ball, R.M. (2002) Re use potential and vacant industrial premises: revisiting the regeneration issue in Stoke on Trent, *Journal of Property Research*, 19, 93–110.

Baum, A. (1991) *Property investment depreciation and obsolescence*, London: Routledge.

Binder, M. (2003) Adaptive reuse and sustainable design: a holistic approach for abandoned industrial buildings, Master's Thesis, University of Cincinnati, OH.

Blakstad, S.H. (2001) A strategic approach to adaptability in office buildings, PhD Thesis, Norwegian University of Science & Technology, Trondheim, Norway.

Bottom, C. W., McGreal, W. S. Heaney, G. (1999). "Appraising the functional performance attributes of office buildings." *Journal of Property Research* 16(4): 339–358.

Building Research Board. (1993) *The fourth dimension in building: strategies for minimizing obsolescence*, Washington, DC: National Academy Press.

Bullen, P.A. (2007) Adaptive reuse and sustainability of commercial buildings, *Facilities*, 25(1–2), 20–31.

Chandler, I. (1991) *Repair and refurbishment of modern buildings*, London: B.T. Batsford Ltd.

Colchimario, R. (2006) Government offices: condo conversions rise up in downtown Birmingham, *Commercial Property News*, 20(3), 43.

Commonwealth of Australia. (1992) *The national strategy for ecologically sustainable development*, Canberra: Australian Government Publishing Service.

Cooper, I. (2001) Post occupancy evaluation: where are you?, *Building Research and Information*, 29(2), 158–163.

Cowan, P., Nutt, B., Sears, D. and Rawson, L. (1970a) *Obsolescence in the built environment: a feasibility study*, London: Joint Unit for Planning Research.

Cowan, P., Nutt, B., Sears, D. and Rawson, L. (1970b) *Obsolescence in the built environment: some concepts of obsolescence*, London: Joint Unit for Planning Research.

Cowan, P., Nutt, B., Sears, D. and Rawson, L. (1970c) *Obsolescence in the built environment: some empirical studies*, London: Joint Unit for Planning Research.

Davis, Langdon (2008) Opportunities for existing buildings: deep emission cuts, *Innovative Thinking*, 8.

Department of Environment and Heritage (2004), Adaptive Reuse, Commonwealth of Australia, Department of Environment and Heritage, Canberra.

Department of Premier and Cabinet. (2008) Growing and linking all of Victoria. Available at http://www.dpc.vic.gov.au/. Accessed on 4 November 2012.

Douglas, J. (1994) Developments in appraising the total performance of buildings, *Structural Survey*, 12(6), 10–15.

Douglas, J. (2006) *Building adaptation* (2nd edition), London: Elsevier.

DTZ. (2012) *Nederland Compleet* 2012, D. Z. v.o.f. Amsterdam, DTZ Zadelhoff v.o.f.

Egbu, C.O. (1997) Refurbishment management: challenges and opportunities, *Building Research and Information*, 25(6), 338–347.

EKOS Limited and Ryden Property Consultants. (2001) *Obsolete commercial and industrial buildings*, Edinburgh: Scottish Executive Central Research Unit.

Elkington, J. (1997) *Cannibals with forks: the triple bottom line of 21st century business*, Oxford: Capstone.

Ellison, L. and Sayce, S. (2007) Assessing sustainability in the existing commercial property stock, *Property Management*, 25(3), 287–304.

Foran, B. and Poldy, F. (2002) Future Dilemmas: Options for 2050 for Australia's Population, Technology, Resources and Environment, Report to the Department of Immigration and Multi-cultural and Indigenous Affairs, Canberra: CSIRO.

Gardner, J.E. (1989) Decision-making for sustainable development: selected approaches to environmental assessment and management, *Environment Impact Assessment Review*, 9(4), 337–366.

Garnaut, R. (2008) *The Garnaut climate change review: final report*, Melbourne: Cambridge University Press.

Goodier, C. and Gibb, A. (2004) *The value of the UK market for offsite*, Loughborough: Buildoffsite and Loughborough University.

Goodland, R. (1995) The concept of environmental sustainability, *Annual Review of Ecology and Systematics*, 26, 275–304.

Heath, T. (2001) Adaptive re-use of offices for residential use: the experiences of London and Toronto, *Cities*, 18(3), 173–184.

Highfield, D. (2000) *Refurbishment and upgrading of buildings*, London: E & FN Spon.

Hill, R. and Bowen, P. (1997) Sustainable construction principles and a framework for attainment, *Construction Management and Economics*, 15(3), 223–239.

Jarchow, S.P. ed. (1991) *Graaskamp on real estate*, Washington, DC: Urban Land Institute.

Johnstone, I.M. (1995) An actuarial model of rehabilitation versus new construction of housing, *Journal of Property Finance*, 6(3), 7–26.

Jones Lang LaSalle (2008) Sustainability now an investor priority, press release, Jones Lang LaSalle, Sydney, Australia.

Kelly, M.J. (2008) Britain's building stock: a carbon challenge. Available at http://www.lcmp.eng.cam.ac.uk/wp-content/uploads/081012_kelly.pdf. Accessed on 11 August 2013.

Khalid, G. (1994) Obsolescence in hedonic price estimation of the financial impact of commercial office buildings: the case of Kuala Lumpur, *Construction Management and Economics*, 12, 37–44.

Kibert, C. (2005) *Sustainable construction: green building design and delivery*, Hoboken, NJ: Wiley.

Kincaid, D. (2000) Adaptability potentials for buildings and infrastructure in sustainable cities, *Facilities*, 18(3), 155–161.

Kincaid, D. (2002) *Adapting buildings for changing uses: guidelines for change of use refurbishment*, London: Spon Press.

Knott, J. (2007) Green refurbishments: where to next? *RICS Oceania e-News Sustainable*, 5–7.

Luecke, R. (2006) *Decision-making: five steps to better results*, Boston, MA: Harvard Business School Press.

Mansfield, J.R. (2002) What's in a name? complexities in the definition of 'refurbishment', *Property Management*, 20(1), 23–30.

Markus, A.M., ed. (1979) *Building conversion and rehabilitation: designing for change in building use*, London: Newnes-Butterworth.

Nahkies, B. (2002) Heritage protection: redefining highest and best use? in proceedings of the PRRES Conference, Christchurch, New Zealand, January 21–23.

Newell, G. and Sieraki, K. (2009) *Global trends in real estate finance*, Ames, IA: Wiley-Blackwell.

Nutt, B; Walker, B; Holliday, S; & Sears, D. 1976. Obsolescence in housing. Theory and applications. Saxon house / Lexington books. Farnborough, Lexington.

Ohemeng, F. (1996) The application of multi-attribute theory to building rehabilitation versus redevelopment options, in proceedings of COBRA, RICS, Las Vegas, NV.

Olson, K. (2005) Windy city conversions: office space goes residential, *Commercial Property News*, 19(14), 30.

PCA (2013) National office vacancy rate down: research. Available at http://www.propertyoz.com.au/Article/NewsDetail.aspx?id=4591. Accessed on 11 August 2013.

Remøy, H. and Van der Voordt, T.J.M. (2006) A new life: transformation of vacant office buildings into housing, in proceedings of CIBW70 Trondheim International Symposium, Norwegian University of Science and Technology, Trondheim, Norway, June 12–14.

RICS (2009) *RICS global commercial property survey Q4 2009*, London: RICS.

Romain, M. (2008) The race begins, *Property Australia*, 23, 24–27.

Savills (2009) Savills research spotlight on Melbourne CBD: November 2009, Available at: http://pdf.savills.asia/australia-office-/savills-research-melbourne-cbd-office-spotlight-2009-q3.pdf. Accessed on 4th December 2010.

Solow, R. (1993) An almost practical step to sustainability, *Resources Policy*, 19(3), 162–172.

Stern, N. (2006) The Stern review: the economics of climate change, HMSO. Available at http://www.hm-treasury.gov.uk/. Accessed on 8 August 2013.

Tosh, D.S. and Rayburn, W.B. (2002) *The uniform standards of professional appraisal practice: applying to standards* (11th edition) Chicago: Dearborn Financial Publishing Inc.

Turban, E., Aronson, J.E. and Liang, T.P. (2005) *Decision support systems and intelligent systems*, Upper Saddle River, NJ: Pearson Education Inc.

UNEP (2006) Sustainable building and construction, United Nation Environment Program. Available at http://www.unep.or.jp/ietc/Activities/Urban/sustainable_bldg_const.asp. Accessed on 11 August 2013.

WCED. (1987) *Our common future*, Oxford: Oxford University Press.

2

Drivers and Barriers for Adaptation

2.1 Introduction

This chapter explains life cycle theory and how it links with adaptation before describing building performance and adaptation theory. From this point drivers and barriers affecting adaptation are detailed. The overarching social, environmental and economic factors are explained as a precursor to a discussion on the specific building attributes associated with adaptations. In this way a comprehensive overview of the theoretical framework in which adaptation decisions are made is provided.

2.2 Building Life Cycle Theory

The concept of life cycle is that there is a beginning, middle and an end; all organisms experience life cycles of varying lengths and buildings are the same. The theory is applied to costs and allows practitioners and researchers to evaluate the total costs associated with building construction and operation over an expected life cycle term. Seven layers of change over time were identified within buildings: the site, structure, skin (building envelope), services, space plan (interior layout), stuff (furniture and equipment) and souls (people) (Brand 1994). There is a sliding scale in terms of the time frames before change occurs. While the site is permanent, the structure lasts from 30 to 300 years, the skin lasts for 20 plus years, services last for 7–20 years, the space plan lasts for 5–7 years, stuff lasts for less than 3 years and the souls change daily (Brand 1994). All buildings contain embodied energy or embodied carbon; that is the energy or carbon emissions that arise from extraction of the raw materials plus assembly into building components

Sustainable Building Adaptation: Innovations in Decision-Making, First Edition.
Sara J. Wilkinson, Hilde Remøy and Craig Langston.
© 2014 John Wiley & Sons, Ltd. Published 2014 by John Wiley & Sons, Ltd.

and transportation to the site. Clearly the longer the life cycle, the lower the total whole life cycle embodied carbon.

What is a typical life cycle for an office building? Estimates vary because of fluctuating conditions and expectations in different property markets globally. For example, an assessment of the Norwegian office market stated that commercial building structures have a usual life cycle of 50 years (Arge 2005). Typically within the 50-year time frame, the building's services will need to be replaced and upgraded three times, due to improvements in technology and increases in user expectations. In US or UK markets, the typical life cycle for commercial buildings differs. The space plan element will be changed the most frequently, typically every 5–7 years, though often less. In Australia, lease terms for commercial buildings are usually 5 years, and therefore the fit-out will change more frequently than in markets where the norm for lease terms is much longer as say in the UK. This is another important factor that, of course, has a significant impact in the whole life sustainability of the building.

Each element of a building has a typical life cycle. The building structure should last 80–100 years plus, the envelope or skin typically lasts for 60 years or so, services 20–30 years and the interior fit-out 5–10 years (Duffy cited in Brand 1994). Services often represent a substantial proportion of construction costs. Given that the age of Melbourne central business district (CBD) commercial stock is on average approximately 31 years old, most buildings would need a service upgrade which is an opportunity to increase the operational sustainability of the building (Jones Lang LaSalle 2008). Similar age profiles for commercial office building stock exist in Sydney; however, in other European markets, like London, an older age profile is apparent. Over the whole building life cycle, most expenditure and environmental impact occurs during the operational phase of the life cycle. Additionally the economic impact of rising energy and other operating costs has increased significantly over the last three decades (Romain 2008). The need to focus on existing stock is a conclusion many have reached, and in 2008 approximately 71% of Australian investment was used for upgrading and building maintenance (DEWHA 2008; PCA 2008a). This figure indicates the significance of the adaptation sector, where the total value of the PCA/IPD Australian Property Index is 121.4 billion Australian dollars as of March 2011 and covers 1535 investments (IPD 2011).

Initially the total building costs are proportioned fairly evenly with the structure costing slightly more than the services and space plan. This represents the traditional view of building costs that takes account of the initial costs and does not consider the ongoing or life cycle costs of buildings. Over time the expenditure on the services and the space plan mean that at the 50-year point, the total costs are highest for the space plan followed by the services (Duffy and Henney 1989). The structure costs are significantly lower at this point in time. The analogy of theatres has been used to describe the notion that a building needs to have adaptability designed in so that it can be altered easily for future changes to the service and space plan factors (Arge 2005). In theatres, buildings are required to adapt to the needs of current productions, and over time theatres remain little changed, while the sets and arrangements change regularly to accommodate the plays (Arge 2005).

Changes occur within building life cycles and Douglas (2006) adopted a five-stage cycle. The first stage was labelled 'birth' when a new activity or process is housed by the building and a new user is accommodated. 'Expansion' is stage two where new requirements are accommodated, new services are introduced and the internal layout is adapted. In addition, there is a strain placed on the building fabric, where possible extension may occur and changes in function or spatial performance may result. 'Maturity' is the third stage, where either uses continue to fit the building and periodic maintenance and minor adjustments are made or current needs exceed capacity and new space is taken elsewhere. Stage four is 'redundancy' due to changes in sources of power, societal cultural values, market needs, technology and/or catchment areas: here the building is partially or totally obsolete and may be partly or totally vacant. The building may be subject to vandalism or occupied by squatters, or it may be mothballed or partially or totally demolished. The final stage is 'rebirth' or 'demolition' where thought will be given to reuse and the building restored, refurbished or demolished. At this point the building can be made more sustainable or a new building may be provided. In this concept of life cycles, adaptation can take place at every stage after 'birth' (Douglas 2006). The level or type of adaptation can and does change according the stage within the life cycle. Minor adaptations give way to more major adaptations over time, and the building meets user needs and the market to a lesser extent. Of course life cycles are closely related to building obsolescence and the issue is covered in detail in Chapter 5.

2.3   Building Performance Theory

Since the 1970s work was undertaken to develop best practice and define building performance theory. Building performance evaluation (BPE) is the process of managed, structured and systematic assessment of building performance in areas such as the structure and fabric and services. BPE sits within a cyclical notion of a building's life cycle. The theoretical framework for BPE evolved out of post-occupancy evaluation (POE). POE is the structured collection of quantitative and qualitative data from building facility managers and users of the building performance. BPE occurs at all stages of the life cycle, whereas POE is undertaken after commissioning of services and initial completion and occupation of the building. According to Preiser (2005), adaptation takes place at the end of the useful building life cycle or at the point where continued current use is no longer perceived to be economically viable (Preiser 2005). For example, Victoria Brewery in Melbourne was adapted to retail and residential apartments after the existing brewery became unprofitable and the site was sold (see Figure 2.1).

Figure 2.1   Former Victoria Brewery site, Melbourne.

Robust, structured and meaningful methods of building appraisal and evaluation have been developed. Owners and consultants can opt for 'off the shelf' evaluation tools, custom made, or adapt existing tools to suit their needs. It is the extension and evolution of these BPE tools that researchers in building adaptation seek to achieve. The goal is to replicate some of the best practice approaches and strengths of the BPE tools while avoiding the weaknesses. A limitation of some BPE techniques is that they tell appraisers the 'what' of BPE but do not extend to decision-making tools.

2.4   Building Adaptation Theory and Sustainability

The arguments for and against building adaptation are categorised broadly under the headings social, economic and environmental. In addition, there are regulatory and legal, location and site, and physical factors which affect adaptation, and the notion of sustainability sits within these factors to varying degrees. Figure 2.2 shows a model of the factors that have been identified as influencing the decision to undertake adaptation. All factors have a direct relationship with adaptation; however, some factors have links with other factors (shown as the dotted line in Figure 2.2).

In addition, many studies have identified attribute(s) that makes a building adaptation 'successful', though the concept of 'successful' varies. The categories of attributes typically identified as relating to 'successful'

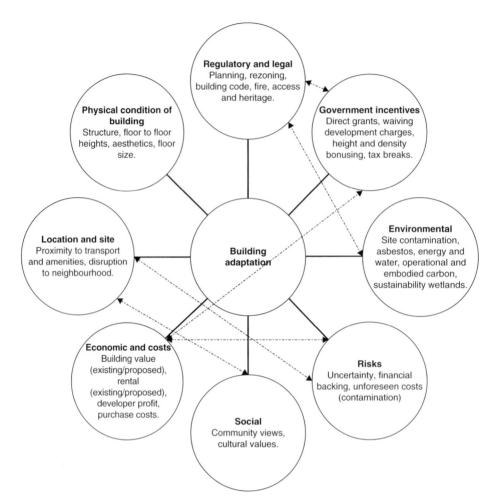

Figure 2.2   Model of decision-making factors in building adaptation.

adaptation are economic, physical, location and land use, legal, social and environmental – as shown in Table 2.1. This next section describes these factors and attributes.

2.4.1   Social Factors

A key argument is that adaptation allows society to retain the social and cultural capital embodied in buildings (Bromley et al. 2005; Bullen 2007). Social sustainability is a fundamental component of Ellington's triple bottom line (TBL) theory (Elkington 1997), and this part of the chapter illustrates both positive and negative aspects and demonstrates that this is another layer of complexity in adaptation.

There are many examples of buildings where the original use has evolved. For example, the Tower of London in the UK has, during its thousand-year

Table 2.1 Building adaptation attributes grouped into categories.

Category	Attribute
Economic attributes	Current value
	Investment value
	Yields
	Increase in value post-adaptation
	Construction and development costs
	Convertibility (ease of conversion to other use and costs associated with the conversion)
Physical attributes	Building height/number of storeys
	Floor plate size
	Shape of floor plate
	Service core location
	Elasticity (ability to extend laterally or vertically)
	Degree of attachment to other buildings
	Access to building
	Height of floors
	Structure
	Floor strength
	Distance between columns
	Frame
	Deconstruction (safe, efficient and speedy)
	Expandability (volume and capacity)
	Flexibility (space planning)
	Technological and convertibility
	Dis-aggregability (reusability/recyclability)
Location and land use attributes	Transport
	Access (proximity to airports, motorways, train stations, public transport nodes, buses and trams)
	Land uses (commercial, residential, retail and industrial or mixed use such as office and retail)
	Existing planning zones
	Rezoning potential
	Density of occupation
Legal attributes	Ownership – tenure
	Occupation – multiple or single tenants
	Building codes
	Fire codes
	Access acts
	Health and safety issues
	Convertibility
Social attributes	Community benefits – historic listing
	Transport noise
	Retention of cultural past
	Urban regeneration
	Aesthetics
	Provision of additional facilities/amenities
	Proximity to hostile factors
	Stigma
Environmental attributes	Internal air quality
	Internal environment quality
	Existence of hazardous materials (asbestos)
	Sustainability issues

history, been used as a prison, a residence and a tourist attraction. It is argued that many generations past, present and future have enjoyed or will enjoy learning about history through experiencing this building, a tangible link to the past with its many rich layers of historical and cultural experiences. Countless other examples of buildings with cultural and historic values exist around the world.

While the retention of cultural and social value is true for some property, it is not supported for all buildings such as those with very poor building quality or those that have a stigma associated with a previous land use. Some form of selection process is required to determine which buildings should be retained for the wider benefit of society and culture. Using Melbourne as an example, the historic listing process provides a method of determining which buildings are considered to have cultural and social significance. The Victorian Heritage Act 1995 administered by Heritage Victoria is the Victorian Government's main form of cultural heritage legislation. The Act enables the identification and protection of heritage places and objects that are of significance to the State of Victoria. The Heritage Act establishes the Victorian Heritage Register, the Heritage Inventory and the Heritage Council of Victoria. However, this listing and protection process is not without flaws; previously some land uses have been excluded from historic listing as they were felt to have little or no value. Industrial buildings are a good example of this omission. Some industrial buildings are deemed to have cultural and social value and are now included in historic listings. However, many industrial buildings we would consider now to have cultural, social and heritage value were demolished and are lost forever. Other city authorities globally have similar legislation to protect the historic built environment. There is another view that some planning legislation relating to heritage and to non-heritage stock limits the scope and extent of adaptation that can be undertaken and in the process compromises the needs of contemporary owners and users (Douglas 2006).

Historic listing is a means of protecting architecturally or socially significant buildings for the wider benefit of society (Ball 2002), and this is a component of social sustainability. A US study concurred with this view of social and cultural worth when researching adaptation of culturally significant industrial buildings (Snyder 2005), though adaptation of heritage buildings can be more expensive due to the additional costs of using traditional building materials, techniques and craftspeople (Bullen 2007). Historic listing is categorised under legal issues as well as social factors.

More broadly, adaptation is part of urban regeneration with aspirations that future generations gain from the protection of buildings (Bullen 2007). Adaptation of buildings within urban regeneration projects delivers social goals such as affordable (or social) housing or employment opportunities in areas of high unemployment (Ball 2002); thus, all sectors of society benefit from adaptation and urban regeneration, not just private businesses. On this basis community views are part of decision-making within planning legislation where changes are proposed to existing buildings as shown in Figure 2.2. In Australia adaptation and regeneration are an essential component of sustainable development facilitating a glimpse of the past, lending character and identity to an area and providing footnotes to history; it is a positive perspective without negative connotations (Australian Government, Department of the Environment and Heritage 2004). There is potential danger that cities face periods where large numbers of obsolete buildings awaiting adaptation blight the region socially (Bryson 1997) and would constitute negative social sustainability. Detroit in the US is a contemporary example.

There are benefits of proactive policies and/or legislation in building adaptation (Highfield 2000; Heath 2001; Ball 2002; Snyder 2005; Burby et al. 2006; Galvan 2006; Shipley et al. 2006). A 2001 study into the adaptation of office buildings into residential buildings in Toronto and London found adaptation rates were higher in Toronto due to a proactive planning policy (Heath 2001). Urban regeneration studies in the London and Bristol docklands found proactive policy and legislation enhanced the retention of existing building stock (Bromley et al. 2005). A proactive policy for adaptation in the New Jersey building codes resulted in an increase in the amount and scope of adaptation in that jurisdiction and demonstrates the way legislation can influence adaptation for the benefit of the community (Burby et al. 2006). The study examined whether code compliance adversely affected existing building adaptation and found that the subcode introduced by the New Jersey authority was statistically significant in its impact on the number of residential adaptation projects and that there was a positive effect on the attitudes towards legislation (Burby et al. 2006). The building codes pre-1994 and post-2002 were compared to evaluate whether a difference existed and examined 117 jurisdictions (Burby et al. 2006).

A social argument against adaptation is the standards required by contemporary buildings, and users are not achievable by adapting existing stock, where in some cases indoor air quality, thermal and acoustic performance requirements cannot be met (Bullen 2007). User expectations of building quality, especially internal environment quality, rise over time (Pinder et al. 2003). Compliance with performance standards varies depending on the physical form of the building to be adapted and the end use required. However, the argument that users occupy substandard space with adapted buildings is strongly countered by the extensive number and wide range of building types successfully adapted and reused (Bullen 2007).

A barrier for adaptation is that the creative component of new build is absent, though the creativity lies in fitting contemporary needs into the old and not starting with a blank canvas (Bullen 2007). An extension of this argument is that commercial buildings of the 1950s and 1960s were too 'ugly' to be retained and adapted (Bullen 2007). There are two contrary views that can be posited. Firstly, the perceptions of the aesthetic qualities of buildings from the 1950s and 1960s are subjective; clearly the buildings were deemed aesthetically appropriate at the time because planning permission was awarded. It is possible that these buildings will be appreciated in the future. Secondly, it is possible to substantially modify a building's external appearance with adaptation.

Another argument is that social goals are not always realised. Examining the social aims of three major UK urban regeneration projects, London Docklands, Cardiff Bay and the Bristol Maritime Quarter, all failed to produce the social mix in the project goals, though the economic goals were achieved (Bromley et al. 2005). Each project involved building adaptation as part of a wider regeneration of areas suffering social and economic blight. Clearly if the social goals of these high-profile projects cannot be achieved, stakeholders should be wary of projects promising significant positive social impact (Bromley et al. 2005). Furthermore, social goals embody a large part of the social sustainability agenda such as creation of employment opportunity within the area or the provision of affordable housing for local people.

Some buildings have or are located near to the so-called hostile factors that can adversely affect a project. Hostile factors include noise pollution, proximity to a noisy motorway or air traffic noise; such environments tend to be less desirable for people. A further category is the presence of deleterious materials such as asbestos. The presence of these materials presents a health hazard to users and occupiers, and remediation and removal costs are high. Further examples with regard to health and sustainability are the presence of volatile organic compounds in building materials such as formaldehydes in glues that emit gas and can cause allergic reactions in occupants and users. The presence of lead in pipework can erode in soft water and be ingested by occupants causing cancers. Another building-related illness (BRI) is caused when legionella bacteria migrate from wet cooling towers associated with building air conditioning and infect occupants with 'legionnaire's disease'. BRIs are those which occur as a result of an individual being in a particular building; sick building syndrome (SBS) is a set of conditions such as eye irritation and dry throats which occur when an individual is inside a building and disappear when the user leaves the premises. In summary, hostile factors can present social and economic barriers that drive up costs to a point where adaptation becomes uneconomical (Bullen 2007).

A supplementary dimension of the social argument derives from stigma associated with previous use that makes buildings unsuitable for adaptation (Kucik 2004). Not surprisingly, the concept of stigma varies between countries, for example, in the US asylums or prison buildings were considered unlikely to be attractive propositions for adaptation to residential (Kucik 2004). However, in Australia, there are examples of successful building adaptation of prisons for residential use, such as Pentridge Prison in Coburg, Melbourne, which housed notorious prisoners such as nineteenth-century bushranger Ned Kelly. Given Australia's convict history, the social status and/or perceptions of prisons may be different to those of other countries. Adaptation of churches and places of worship presents some cultural and social issues in different countries and is another example of the suitability of some land uses for adaptation. Thus, it appears that in some cases stigma issues will have a negative effect on the decision to adapt a building and that this aspect is very much influenced by the local social, cultural and traditional beliefs and conditions. Change of use adaptations of mental asylums are problematic in many countries, although general hospital conversions to residential land use are considered acceptable and many examples exist in the UK and Australia.

2.4.2   Environmental Factors

There is a very strong argument especially with regard to sustainability. The most significant environmental impact of buildings is the greenhouse gas (GHG) emissions associated with energy use (Douglas 2006). The contention is adaptation is inherently sustainable because it involves less material use (i.e. resource consumption), less transport energy, less energy consumption and less pollution during construction (Johnstone 1995; Bullen 2007). Furthermore, embodied energy contained within existing stock is considerable, and the Australian Greenhouse Office estimated the reuse of building materials saves approximately 95% of embodied energy (Binder 2003). Embodied energy is the energy used in the original construction of the building, and this energy is lost when the building is demolished and nothing is salvaged for reuse and/or recycling because all materials are sent to landfill sites. Demolition is wasteful in terms of materials (Australian Government, Department of the Environment and Heritage 2004). The environmental argument is so convincing that a UK study noted even when economic costs are high, environmental and social arguments can sway the decision in favour of adaptation (Ball 2002).

Adaptation of existing stock and the notion of recycling buildings are the most critical aspects of improving sustainability of the built environment (Cooper 2001). The negative impacts of buildings on the natural environment are degradation of habitats, altered ecosystems and reductions in biodiversity resulting from land use. Furthermore, impacts such as reductions in air and water quality contribute to the emergence and spread of infectious diseases that affect humans as well as animals (Koren and Butler 2006). Adaptation presents the opportunity to integrate sustainability retrospectively, and thus the environmental argument is strong (Langston 2010).

Environmental assessment tools were developed to assess and measure building impact on the environment. Buildings assessed under recognised environmental assessment methods such as the UK's Building Research Establishment Environmental Assessment Method (BREEAM) or Leadership in Energy and Environmental Design (LEED) in the US or *Green Star* in Australia have met specified standards in respect of a range of sustainability criteria including energy use. The BREEAM for new office buildings was established in 1990, with a version for existing offices following in 1994. BREEAM aimed to set a benchmark and a framework to evaluate the environmental credentials of buildings. Other assessment tools such as *Green Star* and LEED have followed, signifying the perceived importance globally of reusing existing buildings to deliver a more sustainable built environment. Collectively these tools are well represented in many countries as the standard method of environmental assessment for commercial buildings. Some tools are restricted to limited land uses; BREEAM covers a wider range of land use types than *Green Star*, for example. Furthermore, some tools offer versions that cover different phases of use such as design or operation. BREEAM has a refurbishment tool that *Green Star* has not developed to date. It is an important development for building adaptation to have a validated refurbishment tool that is accepted by the market. To date some tools do not include embodied energy in existing buildings offset against the new build alternative, and this is a weakness of these tools but an aspect that is likely to change in time.

In Australia *Green Star* was developed and monitored by the Green Building Council of Australia. The measure of energy consumption and emissions is incorporated into the National Australian Built Environment Rating System or NABERS (Department of Environment, Climate Change and Water 2010b). NABERS is a national initiative for a performance-based rating system for existing buildings, including offices. NABERS ratings for offices include energy, water, waste and indoor environment on a scale from 1 to 5 stars. In 2010 Mandatory Disclosure legislation requires owners of space exceeding 2000 m² to advertise the building's NABERS rating; similar European legislation exists, known as Energy Performance Certificates (EPCs). Such legislation raises awareness and creates a market for energy-efficient space. Buildings that are accredited under these schemes have demonstrated a level of sustainability. *Green Star* covers a range of building types such as retail, education, office (design), office (as built) and office (interiors), with office (existing building), mixed use, healthcare and industrial buildings under pilot scheme development. *Green Star*-rated buildings contain environmental attributes in terms of energy and water consumption, materials specification, waste and recycling and management (Australian Green Building Council 2010). Other countries have similar tools to assess building energy performance and sustainability.

In addition, adaptation in older industrial areas and working class suburbs is a trend that could limit urban sprawl, stabilise the requirement for the use of concrete and other materials and reduce material flows associated with building construction (Douglas 2006). Thereby, adaptation can deliver environmental sustainability as well as social and economic sustainability. Furthermore, adapting building shells with energy-efficient cladding systems to achieve higher energy ratings could stabilise energy use in the residential and commercial building sectors.

The environmental argument against adaptation arises where the building has excessive amounts of deleterious or hazardous materials, such as asbestos, that pose unacceptable risks to human health (BRE 2009). It is argued buildings with a long history of SBS may not be suitable for adaptation on environmental grounds unless designers and consultants are confident that the SBS issues can be addressed. No other environmental arguments against building adaptation were found which reflects the strong environmental case in favour of adapting existing stock. Environmental issues have a link with regulatory issues. For example, the UK Building Regulations and the Building Code of Australia (BCA) sets minimum energy efficiency standards in respect of some adaptations.

There is a need to adapt existing offices to meet 40% cuts in GHG emissions by 2020 to mitigate climate change (Davis Langdon 2008). This report concluded that emission trading would not deliver sufficient reductions and that capital injection or incentives are required to induce building owners to undertake adaptation (Davis Langdon 2008). The benefits for owners or tenants are lower energy costs, reduced impact of future emission trading schemes, reduced emissions, reduced obsolescence, good risk management strategy, more competitive buildings, improved capital value and increased rental growth. Clearly, some of these benefits involve economic and environmental sustainability. Furthermore, investment in energy efficiency of existing buildings has the potential to reduce GHG emissions by 30–35% within 20 years, faster than alternative approaches (Davis Langdon 2008). In Australia there are approximately 130 million square metres of existing buildings, of which offices comprise 16%, emitting 6.6 million tonnes of GHG per annum (Department of Environment, Climate Change and Water 2010a). There is potential to have a significant effect on emission reductions through adaptation (Australian Government, Department of the Environment and Heritage 2004).

Most building stock is old; 85% of Australian offices are over 10 years old, with the average age being 27 years; it is estimated it would take 290 years to regain the embodied energy in new building through its more efficient performance (Davis Langdon 2008). In Melbourne, average performance of offices is 2 stars or less which is poor; hence, energy performance and the potential for improvement are key environmental adaptation criteria (Davis Langdon 2008). Kelly (2008) noted similar issues of high and increasing emissions and poor-quality stock in the UK.

Water consumption is a very important sustainability indicator and features in many global rating tools (Australian Green Building Council 2010; BREEAM 2013). Most stock was constructed with little attention to minimising water consumption. Adaptation is an opportunity to reduce water consumption through the adoption of measures reducing consumption at the point of use, recycling, harvesting rainwater and reusing water, increasing sustainability. This is a good example of regional variability, and the importance, of different sustainability measures; in the UK the issue is often an excess of water with increased pluvial flooding.

Occupier means of transport to journey to the building has an environmental impact (Davis Langdon 2008). Public transport has lower impact and emissions than private car usage, and proximity to public transport is a positive feature included in environmental assessments (Davis Langdon 2008). Conversely car parking on-site is perceived as a negative within environmental rating tools for office buildings (Australian Green Building Council 2010).

Another relevant environmental aspect identified as important in adaptation is acid rain pollution that causes erosion of stone (Bullen 2007). Acid rain pollution is influenced by prevailing wind patterns and affects some countries whereby deposits are carried across national borders and deposited onto building facades during rainfall. This is another example of region-specific issues, for example, European buildings are affected more so than UK ones. Ozone depletion leads to greater solar degradation of building materials and a faster decline in physical condition and has different impacts in different geographical locations (Douglas 2006). Toxins in building materials can cause allergic reactions in people such as eye irritation. A significant impact is resource consumption and depletion; 40% of global resources are consumed by the built environment, and frequent fit-outs of commercial buildings lead to greater whole life cycle environmental impacts. Adaptation can be sustainable provided it is done within reasonable time frames. During construction negative environmental impacts are excessive noise, dust and dirt (Boyd and Jankovic 1993; Ball 2002).

With the increasing interest in sustainability in the built environment, there has been an increase in the scope and extent of environmental aspects of adaptation (Kincaid 2002; Bullen 2007). There is sometimes an overlap with social, economic and location aspects, for example, proximity to public transport provides environmental, location, economic and social benefits. This overlap means that some attributes can be interpreted on multiple levels.

2.4.3   Cost and Economic Factors

The economics of adaptation are a starting point for many owners considering adaptation and have strong links with risks and government incentives in the decision to adapt as shown in Figure 2.2. Elkington's (1997) third component of sustainability is economic sustainability. Adaptation has to be economically viable to be successful, although economic costs can be traded off against social and environmental gains (Kincaid 2002; Kersting 2006). This perspective comes from TBL accounting theory that has developed with the increased importance of sustainability. In 2007 the UN ratified the standards for urban and community accounting. In TBL accounting environmental issues are taken into account along with social and economic factors.

A compelling economic argument is it is often cheaper to adapt a building rather than demolish or build new (Highfield 2000; Douglas 2006). A study in Stoke-on-Trent in the UK concluded that adaptation was the cheaper option (Ball 2002), a finding that was supported by a subsequent study in New South Wales where financial savings were found when adaptation was compared to new build projects (Department of the Environment and Heritage 2004). However, where new build is straightforward, construction costs are often lower than adaptation (Bullen 2007). Clearly there are factors that can lead to lower or higher comparative costs, and decision-makers need to be cognisant of this issue.

Another argument is that construction periods are reduced because less or no demolition is undertaken, thereby reducing the financing costs (Highfield 2000). A further positive aspect is that older properties can have higher plot ratios for development that have great appeal to developers. Plot ratio is a measure of how much development is allowed on a particular parcel of land. Properties with high plot ratios work favourably with reuse as higher profits can be delivered because higher densities of development are realised

(Highfield 2000). Recently, however, there has been a reversal in planning density policy in many urban centres globally, where the aim is to increase occupational densities in buildings as part of the drive for increased city centre sustainability. In many cases, older stock with more generous space allocation can be reconfigured to increase occupational density.

There is evidence that adaptation increases property value and this is a strong driver. In an investigation of high-density residential property in Hong Kong, a 9.8% increase in property value compared to identical un-refurbished property in the same area was noted (Chau et al. 2003). Another study separated the impact of adaptation on Hong Kong residential property and found 6.6% improvement in value attributed to building adaptation alone (Yui and Leung 2005; Yau et al. 2008). In commercial stock, the expectation would be for higher rental yield post-adaptation or higher capital value at the point of sale. Chandler (1991) noted owners adapt as a means in increasing rental returns. Another positive economic indicator is to have lower vacancy rates in adapted buildings compared to non-adapted stock, which was the case for adapted stock in the Stoke-on-Trent study (Ball 2002).

However, not all projects are economically positive and adaptation costs can surpass a comparable new build. Where original buildings are complex or have requirements due to listing or legislation, costs are likely to be higher than new build, and stakeholders need to appraise this early in decision-making (Holyoake and Watt 2002). Perceptions of higher construction and project costs can result in adaptations being ruled out; one study found that Canadian bankers and developers 'thought' costs were too high which prevented many investors from considering adaptation (Shipley et al. 2006). Uncertainty and perceived higher risk made it harder for developers to secure financial backing on adaptation projects particularly where site remediation was concerned which led to a reduced number of adaptation projects overall (Shipley et al. 2006). Additional issues with adapting existing buildings are maintenance and costs, installation costs and cost constraints compared to cost savings (Holyoake and Watt 2002).

A different issue is poor build quality that can drive up adaptation costs to the point that new build becomes more viable (Bullen 2007). When indirect costs such as disruption caused by adaptation, loss of tenant goodwill and loss of amenity during the works are added in, the economic argument weakens further (Chau et al. 2003). Loss of tenant goodwill and loss of amenity during works are difficult to quantify absolutely in monetary terms and rely on a degree of subjective opinion to evaluate the overall economic cost. A study into the financial drivers for adaptation in Nottingham, England, found financial grants and incentives were needed to promote adaptation of secondary office space which had been vacant or partially occupied for some time (Bryson 1997). In other words, in certain markets adaptation requires financial incentives or funding to make the business case favourable, the building's physical condition has a critical impact on viability, and indirect costs have to be factored into economic decision-making. This approach has been adopted in Melbourne by the Sustainable Melbourne Fund, which has established a system called environmental upgrade agreements (EUA) whereby owners are able to get discounted loans from lenders for sustainable building adaptations. These loans are paid back through the rates that are paid by tenants, who benefit from lower operating costs that result from the sustainability measures introduced in the adaptation. Thereby, the split incentive problem, whereby one party pays for an improvement which another benefits from, is sidestepped.

Cost is a powerful driver for adaptation. A study of 2250 UK projects in 2005 showed that adaptation costs were around 66% of new build (Douglas 2006). Generally there is a case for adaptation on the basis of cost alone; however, other factors, such as end use value and physical considerations, are included in the decision. The UK adaptation cost study found flats (units) and community centres were most expensive to adapt, followed by churches, factories, hospitals, public houses (hotels), primary schools and offices, estate housing (social housing) and banks (Ball 2002). Bank building adaptation costs ranked lowest because the buildings required the least amount of work to adapt them for an alternate use (say, to a restaurant or retail outlet) which kept the costs down presumably as many occupiers retain original bank building features. Conversely the costs of adapting flats (units) were highest because of the works required to adapt them and the amount of building services required in residential buildings (see Part II).

Each stakeholder has different and sometimes competing motivations with regard to adaptation. In respect of costs there are capital costs of adaptation, the ongoing maintenance costs and operating costs that are paid by different stakeholders. The impact of decisions relating to costs varies depending whether the stakeholder is a user or owner in the commercial market. This is important because developers are generally not concerned with life cycle costs if the project is a 'develop and sell' and focus mainly on capital cost (Wilkinson and Reed 2008). On the other hand, users and tenants are concerned with building operating costs, and owners concentrate on financial return, that is, rental levels and vacancy rates. Finally the community concern is the level of amenity, if any, provided by the building adaptation.

Where yield and value is concerned, a critical aspect is to ascertain the clients intentions regarding the end product, for example, do they wish to sell the adapted building on the open market or do they intend to occupy the building themselves (Swallow 1997)? Based on the two different outcomes, different features may be more or less important within the adaptation and may lead to different investment values and yields. A Norwegian study found that owner-occupied stock had a higher incidence of 'adaptability' criteria compared to speculatively designed office buildings. Criteria associated with ease of adaptation were found more often in owner-occupied buildings, and the conclusion was owner-occupied buildings would be more adaptable for long-term occupation and provided a greater return on investment over the whole building life cycle (Arge 2005).

Clearly there has to be market demand to bring about economically sustainable and viable adaptation. Positive user demand and active marketing by stakeholders were important criteria in successful building adaptation in the reuse of vacant stock in Stoke-on-Trent, England (Ball 2002). In that market, user demand was for vacant industrial premises adapted specifically to provide low-cost accommodation for new start-up businesses (Ball 2002). Market research into demand is an important aspect of economically 'successful' sustainable building adaptation because it affects the economic attributes such as yields, post-adaptation value and investment.

Depending on the condition of the original building, it is possible to increase the overall quality with adaptation (Snyder 2005; Kersting, 2006). Quality is measured in various ways, but generally and across all land uses, it can provide a greater number of amenity features, attributes and/or a higher standard of services, features, fixtures and fittings. In Australia, offices are graded by the Property Council of Australia as 'premium' (the best quality and highest rental levels), A, B, C and D grades. D grade is the lowest office grade with the least level of services and amenity and the lowest rental levels. It is possible to increase the office quality grade from one band to another and increase the rental and capital value of the building. However, the capacity to upgrade an office building from one grade to a higher grade is dependent on the condition and location of the building (Isaacs cited in Baird et al. 1996; Kersting 2006).

A UK study showed that post-adaptation buildings typically had lower running and operating costs than prior to the adaptation (Kincaid 2002). An international comparative study of energy efficiency in refurbishment undertaken in the late 1990s and comparing cases in New York, Amsterdam, London, Toronto and Hamburg found that all projects resulted in energy efficiency even if it had not been a high priority and the gain resulted from improvements in the technological specification of service components (Wilkinson 1997). The lower running costs accrue as a benefit of technological advances in building services since the original installation was made. This reduction in running costs is another economic characteristic that could contribute to higher rental levels or higher capital values and is another positive economic characteristic of adaptation (Kincaid 2002). It also indicates fewer GHG emissions are emitted on a square metre basis post-adaptation and that environmental sustainability is enhanced too.

It is important to know whether the property will be sold or leased post-adaptation: in other words, is the person undertaking the adaptation an owner–occupier or lessee? Owner–occupiers are more likely to undertake larger-scale adaptation than lessees who would be more likely to undertake minor fit-outs tailored specifically to their needs (Swallow 1997). Furthermore, who is the owner? Are they private individuals or is the owner an institutional owner; is the property part of a portfolio of properties managed professionally by a financial institution? Different types of owner are likely to undertake different types and levels of adaptation. Institutional investors buy and sell property in the short, medium and long term for profit and are more likely to use professional consultants to advise them on the market conditions and adaptation potential. Whereas private individuals may or may not have ready and immediate access to professional advice, the tenure or ownership of the building may contribute to the degree and extent of adaptation activity.

2.5   Other Attributes Associated with Adaptation

2.5.1   Physical Attributes

The most prevalent category of building-related attributes is physical characteristics (Table 2.1). Clearly the physical building determines to a large extent whether adaptation is possible. All studies covering a wide range of land uses and countries identified the building age as an important consideration (Nutt et al. 1976; Barras and Clark 1996; Ball 2002; Fianchini 2007).

Some buildings, such as those constructed in the 1960s, feature certain forms of construction and materials, for example, asbestos, which make adaptation more expensive or challenging because of the need to comply with strict legal requirements (Bullen 2007). The use of asbestos in construction, usually as a form of fire protection, creates problems in adaptation due to the difficulties of removal without destroying or compromising the structural integrity of the building. In some countries, and in some circumstances, asbestos left in situ is clearly documented; however, other jurisdictions and building owners require removal.

Building height is important in adaptation (Povall and Eley cited in Markus 1979; Gann and Barlow 1996). The type of construction of the frame and the condition of the frame is important. Many concluded steel-framed buildings were more easily adapted because of the ease of cutting into steel beams compared to concrete structures (Povall and Eley cited in Markus 1979; Gann and Barlow 1996; Kincaid 2002). A US study about integrating the past and present through adaptation noted the frame and construction type as significant factors (Kersting 2006). A frame in sound structural condition has the potential to accommodate adaptation, whereas a frame in poor condition will require extensive costly works to accommodate a new or changed use impacting on economic viability.

Floor size in the London office market is an important characteristic in adaptation (Kincaid 2002). Office buildings with unusually shaped floor plates or sizes are more difficult to adapt than those having large open plan space that is suited to subdivision in a number of ways for different user needs (Kincaid 2002). 30 St Mary Axe London, formerly known as the Swiss Re Tower, with its circular floor plate failed to lease as quickly as other buildings with a similar specification because users found the curved floor plate created unusable space. Buildings with unusual irregular plan shapes are harder to adapt to suit a wider range of new users (Kincaid 2002). Plan shape impacts on adaptability and market appeal.

The location of the service core is important (Gann and Barlow 1996; Snyder 2005; Szarejko and Trocka-Leszczynska 2007). It can be located centrally, offset towards the front or rear of the building or in dual/multiple locations. The location affects the ability to subdivide the space as it affects how services can be delivered to various parts of the building. Depending on the size and shape of the floor plates and whether the demand is for large or smaller floor areas, the location of the service core can affect how easy and costly adaptation is. Often a central location will give greater scope for subdivision of the floor plate while minimising corridor and circulation space.

The site affects the adaptation potential, for example, whether a building is detached or attached on one or more sides affects the ease or desirability for adaptation. With less attachment to other buildings, contractors are able to undertake their operations with greater speed and less disruption to any remaining occupants (Isaacs cited in Baird et al. 1996). Similarly, building access, or the number of entry and exit points, is a vital characteristic affecting adaptation potential across a range of property types (Povall and Eley cited in Markus 1979; Gann and Barlow 1996; Kersting 2006; Remøy and Van der Voordt 2006). The more access points a building has, the more flexibility there is for adaptation.

A European office adaptation study found that an optimum floor-to-floor height of a minimum 3.60 m gross or 2.70 m net existed (Arge 2005). In Australia there will be a preference for certain floor-to-floor heights for building adaptation. While no published information in relation to adapting existing buildings was found, design guides for new buildings state the optimum floor-to-ceiling heights in offices are 3.6 m for ground floors and 2.6 m for upper floors (Ryde 2006). Note that Ryde (2006) measured floor-to-ceiling heights using Australian conventions, whereas Arge (2005) measured floor-to-floor height, which includes the slab. Floor-to-ceiling heights are important as they indicate what building services might be accommodated within ceiling voids or raised floors, with different land uses requiring different ideal heights.

Building width is important; one study established a benchmark for building width in adaptations of 15–17 m (Povall and Eley cited in Markus 1979) substantiated 26 years later by Arge (2005). The studies showed that buildings of certain widths were more adaptable; they were able to accommodate a range of space configurations and user needs more frequently. Similarly the technical grid, or the distance between the structural columns on the floor plate (Table 2.1), within the building affected the ease with which it could be adapted for new and other uses, and an optimum or desirable grid was identified (Arge 2005).

A London study concluded floor strength was important (Kincaid 2002). Buildings with floor strength of 3 kN/m^2 or less suited residential uses; those between 3 and 5 kN/m^2 suited retail, office and hospital uses; those between 5 and 10 kN/m^2 suited light industrial uses; and those buildings with floor strength above 10 kN/m^2 fitted industrial and warehouse uses. Thus, in adaptation, floor strength has to be assessed to determine the land uses that are possible and suited physically to the existing floor structure. For example, it is not possible to accommodate office use in a building where the existing floor strength is 3 kN/m^2 unless strengthening works or replacement of the floor is undertaken.

In regard to office buildings and services, it was found that within the technical grid the type of heating ventilation and air conditioning equipment was significant and whether an allowance had been made to accommodate extra capacity in the original design and specification (Arge 2005).

The provision of raised floors in office buildings allowed for changes to, and upgrading of, cabling for information technology systems to be undertaken with ease. Zone-based Internet communication technology provision allowed for more flexibility and adaptability within office building and was sought after (Arge 2005). Another feature is the suspended ceiling grid where horizontal and vertical soundproof barriers could be fitted to zone off parts of the floor plate for different users. These features were labelled 'generality' (Arge 2005). Of course in contemporary office, much of the computing systems will comprise wireless technology, and this shows how quickly some aspects of adaptation can change in a relatively short period of time.

'Flexibility' is focused on the attributes buildings possess which make them easier to change and adapt (Arge 2005). Modularity, where the building is made up of modules or smaller units that can be rearranged, replaced, combined or interchanged easily, was important. Another vital aspect was termed 'plug-and-play' building elements, which allow for a fast change of layout or change of services and wall systems, for example, office partitions that can be easily dismantled and re-erected to accommodate changes in space plans (Arge 2005). Flat soundproof ceilings allow for an easy change of wall partitions to ensure there were no problems associated with sound transmission following adaptation (Arge 2005). Such an approach has significant implications for enhancing environmental sustainability in adaptation.

The potential for vertical and lateral extension was important in Arge's (2005) study. Buildings with scope for lateral or vertical extension were more adaptable because the overall building size could be increased to suit new uses and occupiers. The study termed this characteristic 'elasticity', and it refers to the ease of extending the building laterally or vertically. Other attributes within elasticity are building form, organisational space and ease of compartmentalisation. Compartmentalisation covers the subdivision of space for different users. Functional organisation of space was important and is the ease with which a change of function can be accommodated. In offices, the provision of fire sprinklers allows for large continuous spaces to be provided where desired and for the building regulations to be complied with. The final component of elasticity within office buildings is the potential for subdivision for either letting or sale purposes. Buildings which can be easily subdivided for sale or lease to a number of different tenants or owners were highly desirable, allowing owner's to keep abreast of changes in market demand for office space (Arge 2005).

2.5.2   Locational and Land Use Attributes

Location affects adaptation (Highfield, 2000; Kincaid, 2002; Douglas, 2006). It can be considered in the context of proximity to public transport an environmentally sustainable aspect in adaptation because it reduces car use and the associated GHG emissions. In contrast, however, the amount of on-site parking provision is deemed important for adaptations where little or no public transport is available (Douglas 2006). This is an example of an attribute that can be both positive and negative depending on context and reflects the complexity of decision-making in adaptation in general and in sustainable adaptation in particular. Each case has a unique combination of attributes.

Previous research showed that land use attributes are important in determining whether adaptation is successful or otherwise (Arge 2005). The existing land use affects the potential for a new or changed land use to some degree, for example, an office to a residential change of use. The existing planning zone determines legally what is considered permissible development in a designated area. In Melbourne the term adopted by the legislation, the Planning and Environment Regulations 1998, is planning zone. The planning zones are coded for particular land uses such as, residential (R1Z), business (B1Z), mixed use (MUZ) or other. The zones are listed in the planning scheme, and each zone has a purpose and set of requirements. It is possible to have sites rezoned, for example, industrial or office zones can be re-classed for residential land uses. A proactive policy from the city authority in Toronto during the 1990s promoted increased adaptations from office use to residential (Heath 2001). Operational land use issues include the density of occupation of the land. Within Australia, as in other countries, there is an ongoing planning land use debate about increasing density of the built environment to prevent erosion of green belt land within and around the existing city (City of Melbourne 2005). Increasingly city authorities have been amenable to increasing density of occupation as part of a strategy for increasing sustainability of the city centre.

2.5.3   Legal Attributes

Buildings have legal issues attached to ownership and transfer of ownership and in regard to leasing. Previous studies concluded that adaptation is affected by tenure and whether the person undertaking the adaptation is the owner or a lessee (Swallow 1997). The period for which the party has an interest in the building is important and affects the funds they are willing to invest in adaptation. For example, an owner has an interest in perpetuity, whereas a lessee's interest will last for the duration of the lease term, typically 5 years in Australia and the US. In other property markets, such as the UK, longer lease terms are typical; however, the trends are for lease duration to decline in Europe.

As mentioned previously, owners can be institutional or private. Institutional owners are defined as pension or superannuation funds, financial companies or organisations that invest in property for stakeholders such as investors. As such institutional owners seek to maximise the return on investment in property or buildings for their stakeholders or shareholders and engage professional building and property consultants to ensure this is achieved. Institutional investors are likely to use professional consultants to advise when adaptation is economically and physically desirable. Private owners may be companies or individuals who may or may not use professional consultants to the same degree as institutional owners. Furthermore, private owners may reside offshore and may hold the property for a number of reasons, for example, for future development or for rental income or capital growth. Finally private owners may engage in less or more adaptation; however, this is unknown at present, as no adaptation studies have recorded such data.

Another legal aspect is the way the building is occupied, for example, some buildings have single tenants, and when the lease expires, the opportunity to adapt arises. However, when a building has multiple tenants, it is unlikely that all leases will expire at the same time; therefore, the building may become partly empty (and not earning income) before all leases have expired and the building is available for adaptation. Alternatively owners can negotiate with tenants to terminate leases early and compensate the tenant or to decant tenants, temporarily or permanently, to other floors within the building while adaptation is undertaken. Temporarily decanting tenants was adopted by the Kador Group in their adaptation of the 28-storey building at 500 Collins Street in Melbourne where three floors at a time were adapted and existing tenants were moved to suit the building programme (Your Building 2009).

The final legal consideration is building classification (Arge 2005). Building codes globally adopt different systems of coding land uses or building types. Under the BCA, each building type has a classification that is given a number to designate its characteristics. Residential buildings are either Class 1 or 10 or high-rise or multiunit residential buildings, Class 2. Office buildings are Class 5 and shops or retail buildings are Class 6 in the BCA. Some buildings have multiple classifications: for example, office and retail or Class 5 and Class 6. Different building standards or regulations apply to the different classes of buildings within the BCA, and where an adaptation involves a change of use from one class to another, different standards will have to be met within certain sections of the BCA. Thus, within Australia, the building class may favour change of use adaptations from one class to others to a greater or lesser extent. For example, changing from residential to office use may be more expensive and complex than a change from office to residential classification under the BCA. In 1996 the BCA changed from a 'deemed to satisfy' code to a performance-based code in line with other countries such as the UK and the US. A performance-based code allows architects, designers and engineers to demonstrate that standards are met through a more flexible way than previously and are submitted as alternate solutions. This change to the BCA could favour increased adaptation because the alternate solutions allow the design team to demonstrate compliance with performance criteria in different ways.

2.6   Conclusion

This chapter explained life cycle theory and how it links with adaptation. It is apparent that a longer life is inherently more sustainable as the carbon investment is spread over a longer time frame. The chapter then described performance theory and how a decline in performance is inevitable over time and leads to a need to adapt. The remainder of the chapter outlined adaptation theory in detail. The complexity of adaptation was illustrated along with the issues relating to sustainability and their relationship to adaptation. The drivers and barriers affecting adaptation were presented and described. The overarching social, environmental and economic factors are explained as a precursor to a detailed discussion on the specific building attributes associated with adaptation. The result is that a comprehensive overview of the framework in which adaptation decisions are made has been provided. This chapter highlights key environmental criteria in adaptation and decision-making. However, sustainability has not been mandated in building legislation globally for a long period. European countries lead the way, for example, the UK introduced energy efficiency in Part L of the Building Regulations in 1984 as a result of oil shortages, miners' strikes and sharp price increases in oil in the 1970s. In Australia, sustainability was mandated in Part J of the BCA as energy efficiency in 2006. The US has lagged behind European nations with some states still to enact energy efficiency legislation in buildings. The factors affecting building adaptation decisions are illustrated conceptually in Figure 2.2; however, this figure omits the significance of the various influencing factors on adaptation decisions, and this is one of the knowledge gaps in this field. Chapter 3 goes on to identify some of the typical measures possible in sustainable adaptation and sets out a model for preliminary assessment of adaptation potential to inform decision-making.

References

Arge, K. (2005) Adaptable office buildings: theory and practice, *Facilities*, 23(3–4), 119–127.

Australian Government, Department of the Environment and Heritage. (2004) *Adaptive reuse: preserving our past, building our future*, Canberra: Commonwealth of Australia, Department of Environment and Heritage. Available at http://www. environment.gov.au/heritage/publications/protecting/pubs/adaptive-reuse.pdf. Accessed on 21 August 2013.

Australian Green Building Council. (2010) Australian Green Building Council website. Available at http://www.gbca.org.au/. Accessed on 2nd September 2013.

Baird, G., Gray, J., Isaacs, N., Kernohan, D. and McIndoe, G. (1996) *Building evaluation techniques*, New York: McGraw-Hill.

Ball, R.M. (2002) Re use potential and vacant industrial premises: revisiting the regeneration issue in Stoke on Trent, *Journal of Property Research*, 19, 93–110.

Barras, R. and Clark, P. (1996) Obsolescence and performance in the Central London office market, *Journal of Property Valuation and Investment*, 14(4), 63–78.

Binder, M. (2003) Adaptive reuse and sustainable design: a holistic approach for abandoned industrial buildings, Master's Thesis, University of Cincinnati, OH.

Boyd, D. and Jankovic, L. (1993) The limits of intelligent office refurbishment, *Journal of Property Management*, 11(2), 102–113.

Brand, S. (1994) *How buildings learn: what happens after they're built*, Harmondsworth: Penguin.

BREEAM. (2013) Building research establishment environmental assessment method: in use. Available at http://www.breeam.org/page.jsp?id=373. Accessed on 11 August 2013.

Bromley R.D.F., Tallon, A.R. and Thomas, C.J. (2005) City centre regeneration through residential development: contributing to sustainability, *Urban Studies*, 42(13), 2407–2429.

Bryson, J.R. (1997) Obsolesence and the process of creative reconstruction, *Urban Studies*, 34(9), 1439–1459.

Bullen, P.A. (2007) Adaptive reuse and sustainability of commercial buildings, *Facilities*, 25(1–2), 20–31.

Burby, R.J., Salvasen, D.A. and Creed, M. (2006) Encouraging residential rehabilitation with building codes: New Jersey's experience, *Journal of the American Planning Association*, 72(2), 183–196.

Chandler, I. (1991) *Repair and refurbishment of modern buildings*, London: B.T. Batsford Ltd.

Chau, K.W., Leung, A.Y.T., Yui, C.Y. and Wong, S.K. (2003) Estimating the value enhancement effects of refurbishment, *Facilities*, 21(1), 13–19.

City of Melbourne. (2005) *City plan 2010: towards a thriving and sustainable city*, Melbourne: City of Melbourne.

Cooper, I. (2001) Post occupancy evaluation: where are you?, *Building Research and Information*, 29(2), 158–163.

Davis, Langdon (2008) Opportunities for existing buildings: deep emission cuts, *Innovative Thinking*, 8.

Department of Environment and Heritage (2004), Adaptive Reuse, Commonwealth of Australia, Department of Environment and Heritage, Canberra.

Department of Environment, Climate Change and Water. (2010a) Available at http://www.environment.gov.nsw.au. Accessed on 6 June 2013.

Department of Environment, Climate Change and Water. (2010b) Available at http://www.nabers.com.au. Accessed on 11 August 2013.

DEWHA. (2008) *Mandatory disclosure of commercial building energy efficiency*, Canberra: Department of the Environment, Water, Heritage and Arts.

Douglas, J. (2006) *Building adaptation* (2nd edition), London: Elsevier.

Duffy, F. and Henney, A. (1989) *The changing city*, London: Bulstrode Press.

Elkington, J. (1997) *Cannibals with forks: the triple bottom line of 21st century business*, Oxford: Capstone.

Fianchini, M. (2007) Fitness for purpose: a performance evaluation methodology for the management of university buildings, *Facilities*, 25(3–4), 137–146.

Galvan, S.C. (2006) Rehabilitating rehab through state building codes, *Yale Law Journal*, 115(7), 1744–1782.

Gann, D.M. and Barlow, J. (1996) Flexibility in building use: the technical feasibility of converting redundant offices into flats, *Construction Management and Economics*, 14(1), 55–66.

Heath, T. (2001) Adaptive re-use of offices for residential use: the experiences of London and Toronto, *Cities*, 18(3), 173–184.

Highfield, D. (2000) *Refurbishment and upgrading of buildings*, London: E & FN Spon.

Holyoake, K. and Watt, D. (2002) The sustainable reuse of historic urban industrial buildings: interim results and discussion, in proceedings of COBRA, Nottingham Trent University, UK, September 5–6.

IPD. (2011) IPD in Australia. Available at http://www1.ipd.com/Pages/default.aspx. Accessed on 4 December 2012.

Johnstone, I.M. (1995) An actuarial model of rehabilitation versus new construction of housing, *Journal of Property Finance*, 6(3), 7–26.

Jones Lang LaSalle (2008) Sustainability now an investor priority, press release, Jones Lang LaSalle, Sydney, Australia.

Kelly, M.J. (2008) Britain's building stock: a carbon challenge. Available at http://www.lcmp.eng.cam.ac.uk/wp-content/uploads/081012_kelly.pdf. Accessed on 11 August 2013.

Kersting, J.M. (2006) Integrating past and present: the story of a building through adaptive reuse, Master's Thesis, University of Cincinnati, OH.

Kincaid, D. (2002) *Adapting buildings for changing uses: guidelines for change of use refurbishment*, London: Spon Press.

Koren, H.S. and Butler, C.D. (2006) The interconnection between the built environment ecology and health, in *Environmental security and environmental management: the role of risk assessment*, Morel, B. and Linkov, I.eds., Dordrecht: Springer, 111–125.

Kucik, L.M. (2004) Restoring life: the adaptive reuse of a sanatorium, Master's Thesis, University of Cincinnati, OH.

Langston, C. (2010) Green adaptive reuse: issues and strategies for the built environment (keynote paper), in proceedings of First International Conference on Sustainable Construction and Risk Management, Chongqing, China, June 12–13, pp. 1165–1173.

Markus, A.M., ed. (1979) *Building conversion and rehabilitation: designing for change in building use*, London: Newnes-Butterworth.

Nutt, B; Walker, B; Holliday, S; & Sears, D. 1976. Obsolescence in housing. Theory and applications. Saxon house / Lexington books. Farnborough, Lexington.

PCA (2008) *The office market report*, Melbourne: Property Council of Australia.

Pinder, J., Price, I., Wilkinson, S. and Demack, S. (2003) A method for evaluating workplace utility, *Property Management*, 21(4), 218–229.

Preiser, W.F.E. (2005) Building performance assessment, from POE to BPE: a personal perspective, *Architectural Science Review*, 48(3), 201–204.

Remøy, H. and Van der Voordt, T.J.M. (2006) A new life: transformation of vacant office buildings into housing, in proceedings of CIBW70 Trondheim International Symposium, Norwegian University of Science and Technology, Trondheim, Norway, June 12–14.

Romain, M. (2008) The race begins, *Property Australia*, 23, 24–27.

Ryde(2006) Development control plan. Available at http://www.ryde.nsw.gov.au/Development/Town+Centres/Macquarie+Park+Corridor. Accessed on 23 August 2013.

Shipley, R., Utz, S. and Parsons, M. (2006) Does adaptive reuse pay? a study of the business of building renovation in Ontario Canada, *International Journal of Heritage Studies*, 12(6), 505–520.

Snyder, G.H.. (2005) Sustainability through adaptive reuse: the conversion of industrial buildings, Master's Thesis, University of Cincinnati, OH.

Swallow, P. (1997) Managing unoccupied buildings and sites, *Structural Survey*, 15(2), 74–79.

Szarejko, W. and Trocka-Leszczynska, E. (2007) Aspect of functionality in modernization of office buildings, *Facilities*, 25(3), 163–170.

Wilkinson, S.J. (1997) Thermal improvements in commercial refurbishment: an international comparison, in proceedings of RICS COBRA, University of Portsmouth, UK, September 1997.

Wilkinson, S.J. and Reed, R. (2008) The business case for incorporating sustainability in office buildings: the adaptive reuse of existing buildings, in proceedings of 14th Annual Pacific Rim Real Estate Society Conference, Kuala Lumpur, Malaysia, January 20–23.

Yau, Y., Chau, K.W., Ho, D.C.W. and Wong, S.K. (2008) An empirical study on the positive externality of building refurbishment, *International Journal of Housing Markets and Analysis*, 1(1), 19–32.

Your Building. (2009) Available at http://www.yourbuilding.org/Article/NewsDetail.aspx?p=83&mid=52. Accessed on 23 August 2013.

Yui, C.Y. and Leung, A.Y.T. (2005) A cost-and-benefit evaluation of housing rehabilitation, *Structural Survey*, 23(2), 138–151.

3

Assessing Adaptation Using PAAM

3.1   Introduction

This chapter presents a conceptual model to frame the multiple attributes and the different levels of adaptation that can occur. The importance of 'green' or sustainable features in previous adaptations is used to develop a decision-making tool for non-experts to undertake a preliminary assessment of adaptation potential. This method for undertaking a preliminary assessment of the potential for adaptations is based on the attributes shown to be most important through analysis of previous adaptations. Finally, a case study is presented to demonstrate the application of the model in practice.

3.2   Preliminary Assessment

This section introduces a model, known as Preliminary Adaptation Assessment Model (PAAM), which allows the early assessment by non-professionals to highlight the potential for building adaptation. PAAM is largely based on Chudley's (1981) decision model modified to accommodate sustainability (see Figure 3.1). The underlying concept is based on the sequence of decisions which have to be undertaken for a robust initial assessment of the suitability of a building for adaptation and the factors outlined in Chapter 2. The PAAM identifies exit points with an indication of the options available to stakeholders at each stage. This initial model sequences factor to reflect a rank order of weighting based on the findings of work conducted by others (Ball 2002; Kincaid 2002; Arge 2005; Remøy and Van der Voordt 2006).

Sustainable Building Adaptation: Innovations in Decision-Making, First Edition.
Sara J. Wilkinson, Hilde Remøy and Craig Langston.
© 2014 John Wiley & Sons, Ltd. Published 2014 by John Wiley & Sons, Ltd.

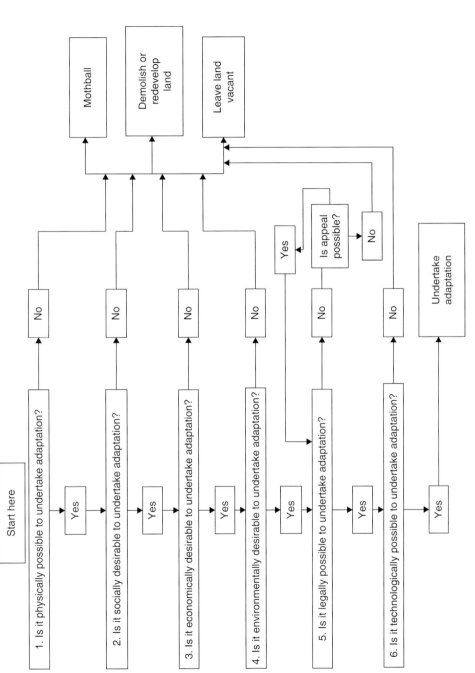

Figure 3.1 Decision-making PAAM for existing buildings.

Given the attributes discussed in Chapter 2, the starting point for the decision is physical suitability; if a building is structurally unsound, there is no possibility of adapting it for further use whatever the regulatory, environmental, economic, technological or social attributes. Physical property attributes featured consistently as very important (Kincaid 2002; Arge 2005; Remøy and Van der Voordt 2006). PAAM acknowledges the broader social and community agenda placing social desirability as the second decision. Economic considerations were placed first in Chudley's (1981) model and, despite strong economic arguments posited by Kincaid (2002), Arge (2005) and others, are placed here after social considerations. Environmental sustainability criteria have gained importance (Bullen 2007; Langston 2007) and are placed fourth. The next step considers regulatory aspects such as planning legislation and building code compliance that were placed second in Chudley (1981). Swallow (1997), Douglas (2006) and Highfield (2000) acknowledged the importance of regulatory aspects of adaptation; however, with the change to performance-based codes, compliance has become a more flexible negotiation than hitherto. Finally, PAAM assesses whether adaptation will deliver a building that meets technological expectations that were omitted from Chudley's (1981) model but identified by Arge (2005) and others (Remøy and Van der Voordt 2007b) as important. If all the stages are satisfied, then adaptation of the building can be undertaken with a reasonable degree of confidence in a 'successful' outcome that meets the needs of stakeholders.

PAAM (Figure 3.1) is predicated on the analysis of multiple criteria in six sequenced stages. Overall, PAAM is a reliable diagrammatic representation of the relationship between key significant decision-making criteria and building adaptation. The model is 'static', analysing data at one point in time. However, PAAM lacks weighting of the stages and quantitative validation of the order of the factors. In the current form there is no quantitative, but some qualitative, evidence to justify sequencing the order of stages. In other words, there is no evidence to support, for example, social considerations being more important than environmental considerations. A further weakness is that the factors are considered as a whole and the importance of the attributes within the factors is unknown. For example, with social factors it is not known which is most important in adaptation – proximity to hostile factors or aesthetics. The weightings presented here (see Table 3.1) were based on the analysis of 1273 'alterations' building adaptations undertaken in the Melbourne CBD from 2009 to 2011 inclusive, and Section 3.3 describes how the weightings were calculated.

Figure 3.2 shows the relationship between the influencing attributes identified in Chapter 2, grouped under board headings social and technological and so on, and building adaptations. Each of the seven categories of influencing attributes may affect building adaptation to some degree or not. Each category will affect the building adaptation, though not all may be present or influence each adaptation to the same extent. In the right-hand section of the conceptual model, building adaptations are grouped into the following outcomes or degrees of adaptation: level 1 is the lowest level of adaptation possible and is labelled 'minor alterations', level 2 is 'alterations',

Table 3.1   Property attributes used in PCA of all 'alteration' adaptations.

1. Building work type	21. Existing land use
2. Total building area	22. Degree of attachment to other
3. Nature of work (redevelopment or	buildings
adaptation)	23. Floor size
4. Occupant classification (owner/lessee/	24. Site access
vacant)	25. Street frontage (m)
5. Plan shape	26. Tenure type
6. Typical floor area	27. Site area (m²)
7. Purpose built for current use	28. Property location
8. Occupancy (sole/multiple/vacant)	29. Historic listing
9. Purpose built commercial	30. *Green Star* rating
10. Zoning	31. Number of storeys (height)
11. Site orientation	32. NABERS rating
12. Gross floor area	33. Age in 2010
13. Aesthetics	34. ABGR rating
14. Net lettable area	35. Year built
15. Internal layout (space plan)	36. Building envelope and cladding
16. Property Council of Australia building	37. Parking
grade	38. Hostile factors
17. Internal layout (columns)	39. Number of car bays
18. Type of construction	40. Elasticity potential – lateral flexibility
19. Vertical service location	41. Proximity to transport
20. Elasticity potential – vertical flexibility	42. Tenure type

level 3 'change of use' adaptation, level 4 'alterations and extensions', level 5 'new-build' works and finally level 6 'demolition' works. So, for example, 'alterations' could comprise the fit-out of floors in an office building and 'alterations and extensions' could comprise the vertical extension of an existing office building through an additional floor. Although the model may appear rigid, the design has included a degree of flexibility and it may be altered as follows: the number of adaptations entered into the model is variable and can be adjusted according to the circumstances surrounding different buildings, cities or locations. In this model adaptations and property attribute variables are examined retrospectively, that is, after adaptation to assess their level of correlation.

In summary, previous adaptations were analysed to identify the important attributes and then to weight the importance of attributes. Finally, the aim was to develop a reliable predictive model for the initial decision on the suitability of a commercial building for adaptation. A building attribute database was compiled comprising all the building adaptations as outlined in building permit data. Previous studies typically adopted a case study approach with detailed analysis of small samples of buildings, and from these studies adaptation criteria were identified (Ohemeng 1996; Blakstad 2001; Heath 2001; Ball 2002; Kincaid 2002; Kucik 2004; Arge 2005; Remøy and Van der Voordt 2007b). The fields in the database were populated with information from multiple sources including 'Cityscope' (RP Data 2008), 'PRISM' produced by the State Government of Victoria's Department

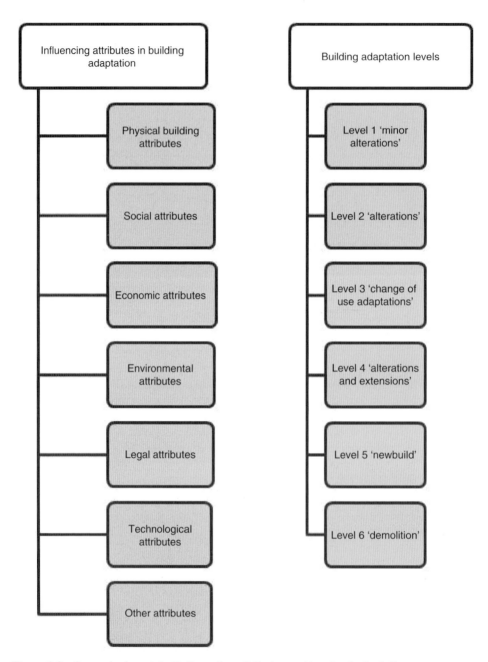

Figure 3.2   Conceptual model of influencing attributes and levels of adaptation.

of Sustainability and Environment (DSE 2008) and the Property Council of Australia (PCA 2007, 2008a). Further data was collected by physically inspecting the exteriors of the properties in the database. Every building adaptation between 2009 and 2011 in the CBD was included in the database and examined. Given the nature of the social, economic, environmental and physical characteristics of the Melbourne CBD building stock, it is likely that the results are applicable to other cities in Australia and internationally. However, the results here and the model are most relevant to Melbourne.

3.3   Principal Component Analysis

Principal component analysis (PCA) is a mathematical technique that converts a set of conceivably correlated variables into a set of values of linearly uncorrelated variables named principal components (Joliffe 2002). The transformation is defined so that the first principal component or factor has the largest possible variance; it accounts for as much variability in the data as possible. PCA condenses information in a number of initial variables into a smaller set of new composite factors with a least loss of information (Hair et al. 1995). PCA is mostly used as a tool for making predictive models. The technique is a reliable, proven method of highlighting dimensions in cross-sectional data to uncover, disentangle and summarise patterns of correlation within a data set (Heikkila 1992; Horvath 1994). All property attributes listed in Table 3.1 were examined to identify the degree of variance explained by an interpretable group of factors.

The analysis produced a smaller number of factors with a final result being a table of identifiable factors that includes the loadings of individual attributes (Table 3.2). 1273 building adaptations were analysed. Assigning meaning involves interpretation of the pattern of the factor loadings, and this is subjective (Hair et al. 1995), based on the coding system derived from the literature. The analysis examined adaptations classed as 'alterations',

Table 3.2 Summary of PCA factors for 'alteration' adaptations.

Factor number	Factor name	Factor attributes
1	Environmental and physical (32%)	Property Council of Australia building quality grade (29%) NABERS rating (26%) Aesthetics (25%) Height (number of storeys) (20%)
2	Social and physical (15%)	Historic listing (42%) Construction type (30%) Parking (28%)
3	Physical (14%)	Street frontage (60%) Vertical service location (40%)
4	Environmental (10%)	*Green Star* rating (100%)

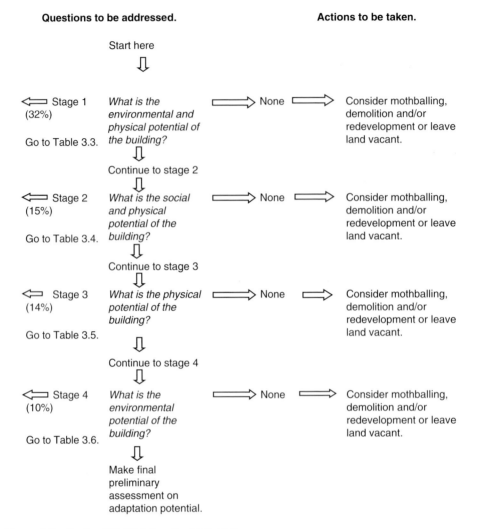

Figure 3.3   Revised PAAM 'alteration' adaptations.

that is, the lesser adaptations (typically fit-outs and minor works). The analysis also showed that ten property attributes accounted for 70% of the original variance and formed the starting point for the revised PAAM. Significantly these ten attributes include two environmental- or sustainability-related attributes: NABERS rating and *Green Star* rating. This result indicates that sustainability-related measures have become statistically important in adaptation. Wilkinson's (2011) earlier study on important attributes in adaptation did not find any environmental attributes to be important, and this follow-up study marks a noteworthy change in this respect. The other attributes were classified as physical and social (based on the categorisation listed in Chapter 2). From the PCA and the results outlined in Table 3.2, it is possible to revisit PAAM in Figure 3.1 and propose a more accurate PAAM in Figure 3.3.

3.4   Preliminary Adaptation Assessment Model

The main objective of PAAM is that it can be used by non-experts to make an initial assessment of a building's general suitability for 'alterations' adaptation. Further detail is shown at each stage with exit points and suggested actions.

What the revised PAAM shown in Figure 3.3 lacks at this point is a quantitative estimation of adaptation potential. It is possible to see that the first factor, physical and environmental, accounts for 32% of importance but it does not indicate which aspects within physical and environmental should be examined. Therefore, the next stage is to use the percentage of occurrence of each attribute within each factor (see Table 3.1) and calculate the most to least important. For example, in factor one, physical and environmental PCA building quality grade accounts for 29% of importance, followed by NABERS rating at 26%, aesthetics at 25% and building height at 20%. For each attribute, it was possible to calculate the percentage of buildings undergoing adaptation with that particular attribute. For each attribute, the percentages of occurrence were standardised between 0 and 1, and a grading score was allocated on a five-point scale which was either very high (0.81–1.00), high (0.61–0.80), medium (0.41–0.60), low (0.21–0.40) or very low (0.00–0.20). For example, for the attribute PCA building quality grade (Table 3.1 and Figure 3.4), the assessor poses the question, 'what is the existing PCA building quality grade?' Based on a univariate statistical analysis of the building attribute

Figure 3.4 Holyman House, Melbourne.

database, the possible answers are unclassified grade (32%), B grade (21%), A grade (20%), premium grade (12%), C grade (11%) or D grade (3%) in order of the per cent occurrence derived from previous adaptations measured in the model. These percentages are standardised and allocated a score as follows:

Percentages: P1 = 0.32, P2 = 0.21, P3 = 0.20, P4 = 0.12, P5 = 0.11, P6 = 0.03

- *Step 1*: Identify highest value, P1: 0.32 = 1.
- *Step 2*: Identify lowest value, P6: 0.03 = 0.
- *Step 3*: Calculate the values for the intermediate answers using the equation:

$$\frac{\text{Value}-\text{Minimum}}{\text{Maximum}-\text{Minimum}} = \frac{0.21-0.04}{0.32-0.04} = \frac{0.17}{0.28} = 0.61$$

The standardised values are P1 = 1, P2 = 0.61, P3 = 0.57, P4 = 0.31, P5 = 0.28, P6 = 0. Based on this approach, the final scores for the PCA building quality grade are based on the grading score and are as follows:

- Ungraded Final score 1 = very high
- A grade Final score 0.61 = high
- B grade Final score 0.57 = medium
- Premium Final score 0.43 = medium
- C grade Final score 0.30 = low
- D grade Final score 0 = very low

This standardisation approach is a useful qualitative aid to scoring the answers where there are more than two options; however, scoring two options is subjective and this is an acknowledged weakness in this approach. The next step is to score each group of graded attributes within a factor and finally the overall building PAAM grading.

Table 3.3, Table 3.4, Table 3.5 and Table 3.6 illustrate how each factor in Table 3.1 is assessed sequentially to determine the potential of a building for adaptation. Table 3.3 poses questions relating to attributes identified in factor one. The first attribute concerns the Property Council of Australia building quality grade and is weighted at 29% of the variance of factor one. The second consideration is NABERS rating that is divided into two categories, either a yes or no outcome. The third attribute is the aesthetic quality that is weighted at 25% of the factor, and buildings are evaluated on the basis of having high to low aesthetic qualities. The next attribute is the number of storeys, weighted at 20% and is assessed on the basis of being low, medium, high or sky rise.

Having answered the questions relating to environmental and physical attributes, the results give an indication whether the building has good potential for adaptation (Table 3.3). After all questions have been answered, it is possible to determine whether to proceed to factor two or consider other options. For example, and based on previous adaptations,

Table 3.3 Predictive model (environmental and physical factor).

What is the environmental and physical potential of the building for adaptation? Start here. Answer each question in turn. When complete, return to Figure 3.3 for the next stage.

Weighting (% of factor variance)	Question	Grading scale (outcomes in rank order)	Notes	Result and grade selected (e.g. very high to very low)
29%	What is the *Property Council of Australia grade* of the building?	1. Ungraded. Very high 2. B grade. High 3. A grade. Medium 4. Premium. Medium 5. C grade. Low 6. D grade. Very low	1. Ungraded (32%) 2. B grade (21%) 3. A grade (20%) 4. Premium (12%) 5. C grade (11%) 6. D grade (3%)	—
26%	Is there a *NABERS rating* of the building?	1. No. Very high 2. Yes. Very low	1. No (54%) 2. Yes (46%)	—
25%	What is the *aesthetic quality of the building*?	1. Attractive. Very high 2. Very attractive. High 3. Neither attractive nor unattractive. Medium 4. Unattractive. Low 5. Very unattractive. Very low	1. Attractive (38%) 2. Very attractive (30%) 3. Neither attractive nor unattractive (17%) 4. Unattractive (13%) 5. Very unattractive (2%)	—
20%	What is the *number of storeys*?	1. 11–20. Very high 2. 41+. Low 3. 1–10. Very low 4. 21–30. Very low 5. 31–40. Very low	1. 11–20 (36%) 2. 41+ (22%) 3. 1–10 (17%) 4. 21–30 (13%) 5. 31–40 (13%)	—
Total 100%				

Table 3.4 Predictive model (social and physical factor).

What is the social and physical potential of the building for adaptation? Start here. Answer each question in turn. When complete, return to Figure 3.3 for the next stage.

Weighting (% of factor variance)	Question	Grading scale (outcomes in rank order)	Notes	Result and grade selected (e.g. very high to very low)
42%	Is there a *historic listing or overlay* relating to the building?	1. No. Very High 2. Yes. Very Low	1. No (74%) 2. Yes (26%)	—
30%	What is the *construction type* used in the building?	1. Concrete frame. Very high 2. Load bearing. Very low 3. Steel frame. Very low	1. Concrete frame (92%) 2. Load-bearing brick or stone (7%) 3. Steel frame (1%)	—
28%	Is there *parking* in the building?	1. Yes. Very high 2. No. Very low	1. Yes (82%) 2. No (18%)	—
Total 100%				—

Table 3.5 Predictive model (physical factor).

What is the physical potential of the building for adaptation? Start here. Answer each question in turn. When complete, return to Figure 3.3 for the next stage.

Weighting (% of factor variance)	Question	Grading scale (outcomes in rank order)	Notes	Result and grade selected (e.g. very high to very low)
60%	What is the *street frontage* of the building?	1. Medium width (21–40 m). Very high 2. Extra wide (61+ m). Low 3. Wide (41–60 m). Very low 4. Narrow (<20 m). Very low	1. Medium (44%) 2. Extra wide (24%) 3. Wide (17%) 4. Narrow (15%)	—
40%	What is the *vertical service location* of the building?	1. Central location. Very high 2. Multiple locations 3. Offset to one side. Very low	1. Central location (51%) 2. Multiple locations (34%) 3. Offset to one side (15%)	—
Total 100%				—

Table 3.6 Predictive model (environmental factor).

What is the environmental and physical potential of the building for adaptation? Start here. Answer each question in turn. When complete, return to Figure 3.3 for the next stage.

Weighting (% of factor variance)	Question	Grading scale (outcomes in rank order)	Notes	Result and grade selected (e.g. very high to very)
100%	Is there a *Green Star rating* for the building?	1. No. Very high 2. Yes. Very low	1. No (98%) 2. Yes (2%)	—
Total 100%				—

if a building has a Property Council of Australia B-grade office, is NABERS rated, is attractive and is 12 storeys high, it has a very high potential for 'alterations' adaptation. Conversely, if the building has less than 11 storeys, has a Property Council of Australia D-grade rating, is not NABERS rated and is very unattractive, it has much lower potential for an 'alterations' adaptation.

Factor two explains 15% of the total variance in 'alterations' adaptations and comprises three attributes labelled 'social and physical' (Table 3.4). The attributes are historic listing, construction type and parking. The first attribute, historic listing, is important and weighted at 42% of factor two. Unlisted buildings were found to be less likely to undergo adaptation. The next attribute, construction type, is weighted at 29% of factor two. In this market concrete-framed buildings were most likely to be adapted, followed by load-bearing and lastly steel-framed buildings. The final attribute in factor two is parking, weighted at 28%. Buildings with parking were much more likely to be adapted than those with no parking on-site.

When all questions are addressed, the responses indicate the suitability for 'alterations' adaptation. By way of an example, an unlisted concrete-framed building with parking scores 'very high' and is likely to undergo adaptation, whereas a listed, steel-framed property without parking is more unlikely to undergo adaptation based on previous adaptations. The next stage is to evaluate factor three attributes that explain 14% of the total variance and contain two attributes labelled 'physical' (Table 3.5). The first attribute, street frontage, is weighted at 60% and is very important, while the second attribute, building vertical service location, is weighted at 40%. Buildings less than 20 m wide are unlikely to undergo work, whereas a property between 21 and 40 m wide is very likely to have good potential for 'alterations' adaptations. With the second attribute, buildings with centrally located service cores have a greater likelihood of adaptation. In summary, for factor three, a building 15 m wide and with services located to the side of the property scores very low, whereas a building 25 m wide with a central location for services scores very high.

The final factor comprises one attribute, *Green Star* rating (Table 3.6). If the building has a *Green Star* rating, it is very unlikely to be adapted. The attribute is very important within this factor.

3.5   Illustrative Case Study

This section discusses and demonstrates the application of PAAM to a real building as an illustrative case study.

3.5.1   Building Description

Holyman House is a three-storey bluestone warehouse constructed in 1858. At some stage a change of use adaptation occurred from warehouse to office land use. The architectural style is Renaissance Revival style, and the building is a typical example of the conservative classical style. Renaissance style is a fifteenth-century revival of classical Rome, where the classical orders were perceived as a foundation for beauty but not a rule to be slavishly adopted. The architectural orders of classicism were applied as conventions for perfect proportion, based on a human scale. In conforming to the general principles of symmetry, geometry and proportion, the Renaissance style is perceived to deliver visually pleasing buildings (see Figure 3.4). Generally Renaissance Revival buildings are limited to around three storeys, as this is considered consistent with human proportions. Holyman House compliments the neighbouring Customs House property and is an integral part of the low streetscape along Flinders Street. The Statement of Cultural Heritage Significance affirms the property is a fine example of a city office building in the Renaissance Revival style, built of bluestone and with details that are severe in outline in keeping with the character of the stone. Holyman House was constructed for Mr. Richard Goldsbrough, a founder of the wool broking firm Goldsbrough, Mort and Company. The building, designed by John Gill, was Goldsbrough's first headquarters and was occupied until 1864 when the firm relocated to Bourke Street.

Holyman House is approximately 786 m² and attached on one side to the rear of the photograph. The vertical services are located centrally and there is no lift. The building is not graded under the Property Council of Australia quality matrix. The building suffers some minor physical obsolescence and may require updating to attract long-term tenants. There are currently two storeys vacant and for lease. Additionally, due to the property's age and condition, it is unlikely that it can fully support a modern office environment, which may lead to the building suffering from technological obsolescence. The internal condition of the property is considered below average to meet the expectations of the majority of tenants. Furthermore, some mild wear and tear is visible throughout the inspected external and internal areas.

3.5.2   Assessing a Building for 'Alterations' Adaptation

A PAAM checklist was used to assess the potential of the building for 'alterations' adaptations. The checklist comprises four sections representing the four factors produced in the PCA. For each factor the assessor records a result for each attribute according to the possible categories offered. For factor one, the first attribute asks, what is the Property Council of Australia building quality grade? From Table 3.3 there are six options to select from and the assessor grade scores the answer where a result of ungraded is scored 'very high'. The results are summarised and presented in Table 3.7 that shows both the data collected for each attribute and the assessment of adaptation potential. Each factor is evaluated and a final overall evaluation is provided at the base of the table.

Within factor one, environmental and physical, there is variation in the scoring within the four attributes from very low (number of storeys) to very high (Property Council of Australia building quality grade). Using the approach described overall environmental and physical potential is ranked as high/very high. Factor two, social and physical potential, shows the building has very low potential for adaptation, with all three attributes scoring very low, and factor two attributes account for 15% of total variance. In factor three, physical potential, the result is very high. Factor four examines the environmental potential for adaptation and a result of very high is noted. Finally, the four factors are assessed to derive the overall potential of the building for 'alterations' adaptation. Given the factor

Table 3.7 Adaptation potential (Holyman House).

Factor	Attribute	Data	Results
1. Environmental and physical potential	a. What is the existing *Property Council of Australia building quality grade*?	Ungraded	Very high
	b. Is there a *NABERS rating* of the building?	No	Very high
	c. What is the *aesthetic quality of* the building?	Very attractive	High
	d. What is the *number of storeys* in the building?	3	Very low
	Environmental and physical result overall	—	**High to very high**
2. Social and physical potential	a. Is there an *historic listing or overlay* relating to the building?	Yes	Very low
	b. What is the *construction type* used in the building?	Load bearing	Very low
	c. Is there *parking* in the building?	No	Very low
	Social and physical result overall	—	**Very low**
3. Physical potential	a. What is the *street frontage* of the building?	23.14 m	Very high
	b. What is the *vertical service location* of the building?	Central	Very high
	Physical result overall	—	**Very high**
4. Environmental potential	a. Is there a *Green Star rating* for the building?	No	Very high
	Environmental result overall	—	**Very high**
	Overall result	—	**High to very high**

grades and the higher loadings of the first and last two groups of attributes, overall Holyman House is considered to have high/very high potential for 'alterations' adaptation.

During the initial investigation of adaptation options, stakeholders, including non-experts, can use the model to determine overall potential, focusing at the outset on the property attributes that account for most variance in adaptation. With this approach, attributes that are not important in adaptation are not considered because the PCA retained important attributes only. Consideration of unimportant attributes is an issue with some decision-making tools (Chudley 1981; Kincaid 2002; Arge 2005; Langston 2007; Remøy and Van der Voordt 2007b; Arup 2008; Bullen and Love 2011). In this respect the model represents an original approach and an important step forward in the decision-making process where adaptation is being considered. A limitation of this approach is that the model looks at adaptation based on what has happened in this market in the period 2009–2011 and it does not look at prevailing market conditions at this level of preliminary decision-making. PAAM facilitates a relatively fast and deeper understanding of the adaptation potential of a building and highlights the important property attributes that are likely to present issues for stakeholders.

The illustrative case study shows PAAM in practice and the considerations undertaken at each stage. For each factor in the predictive model, possible answers are proposed based on the attributes to illustrate how the model determines very high to very low adaptation potential in a building. The model requires an assessment of each attribute within each factor as shown in the case study. The degree of importance of each factor and each attribute within each factor is known. Finally, the model explains 70% of variance in alterations and extension adaptations.

PAAM takes an assessor through a series of stages and significantly does not require them to possess high levels of professional knowledge or technical competence. The assessor is able to use the case study checklist and PAAM to determine the suitability of a building for adaptation based on the empirical analysis of multiple adaptations. At each stage the assessor deliberates only the most important property attributes. For each factor in the model, building profiles are suggested based on the attributes to illustrate how the model can determine high to low adaptation potential. Furthermore, the case study illustrates how the model and checklist work in practice. Following extensive modelling with a number of different quantitative weighting approaches in the predictive models, it was decided that a purely quantitative assessment was not sufficiently robust and reliable.

3.6   Conclusion

There were attributes that previous studies had identified as being important to building adaptation but which were found either to be unimportant in explaining variance in adaptation in Melbourne in the

PCAs or were not possible to test in this study. The first group, unimportant attributes, can be explained partly because of high levels of homogeneity in the stock, for example, the attribute 'proximity to public transport' revealed that because the CBD is so well serviced by public transport, all buildings were located within more or less equal distances to transport services.

For 'alterations' adaptations that are more minor in scope and extent, the key findings derived from the PCA analysis were the following:

1. The Property Council of Australia building grade is the most important attribute.
2. NABERS rating has become an important attribute with buildings with 46% of stock rated.
3. Aesthetics is important and 68% of the adaptations occurred to buildings rated as very attractive and attractive.
4. The building height or number of storeys is an important attribute with 53% of adaptations to stock of 20 floors or less.
5. In factor two historic listing was the most important attribute with 74% of adaptations to unlisted stock.
6. Physical attributes are important in 'alterations' adaptations and include construction type, whereby 92% of adaptations occurred to concrete-framed buildings.
7. 82% of adaptations occurred to buildings with on-site parking provision that was important in factor two.
8. In factor three buildings between 31- and 40-m width were most frequently adapted (31% of the adaptations).
9. Vertical service location was important with 51% of adaptations occurring to buildings with a centrally located service core.
10. In factor four *Green Star* rating was an important attribute whereby 98% of adaptations occurred to buildings without a *Green Star* rating, and this demonstrates the need to upgrade existing stock.

A further limitation of the approach is that the model is derived from an analysis of past practices. In undertaking an assessment, the assessor does not consider current property market and general economic conditions within PAAM; these factors are outside of the scope of the model.

In closing, this approach shows which property attributes are most important in adaptation based on the analysis of all events during an extended time period, and as a result, PAAM is advocated. PAAM provides a decision-making tool for non-experts to make an initial evaluation of potential based on empirical evidence that incorporates sustainability criteria currently found to be important in adaptation. PAAM has been discussed with an illustrative case study to demonstrate how the model might be applied in practice. This approach highlights a method which could lead to more evidence-based decision-making in respect of building adaptation as humankind seeks solutions to the challenges of reducing the environmental footprint of our urban settlements.

References

Arge, K. (2005) Adaptable office buildings: theory and practice, *Facilities*, 23(3–4), 119–127.

Arup(2008) *Existing buildings: survival strategies*, Melbourne: Arup.

Ball, R.M. (2002) Re use potential and vacant industrial premises: revisiting the regeneration issue in Stoke on Trent, *Journal of Property Research*, 19, 93–110.

Blakstad, S.H. (2001) A strategic approach to adaptability in office buildings, PhD Thesis, Norwegian University of Science & Technology, Trondheim, Norway.

Bullen, P.A. (2007) Adaptive reuse and sustainability of commercial buildings, *Facilities*, 25(1–2), 20–31.

Bullen, P.A. and Love, P.E.D. (2011) A new future for the past: a model for adaptive reuse decision-making, *Built Environment Project and Asset Management*, 1(1), 32–44.

Chudley, R. (1981) *The maintenance and adaptation of buildings*, London: Longman Group Limited.

Douglas, J. (2006) *Building adaptation* (2nd edition), London: Elsevier.

DSE. (2008), *PRISM*, Melbourne: Department of Sustainability and Environment.

Hair, J.F., Anderson, R.E., Tatham, R.L. and Black, W.C. (1995) *Multivariate data analysis* (4th edition), Upper Saddle River, NJ: Prentice Hall.

Heath, T. (2001) Adaptive re-use of offices for residential use: the experiences of London and Toronto, *Cities*, 18(3), 173–184.

Heikkila, E.J. (1992) Describing urban structure, *Review of Urban and Regional Development Studies*, 4, 84–101.

Highfield, D. (2000) *Refurbishment and upgrading of buildings*, London: E & FN Spon.

Horvath, R.J. (1994) National development paths 1965–1987: measuring a metaphor, *Environment and Planning A*, 26, 285–305.

Joliffe, I.T. (2002) *Principal component analysis* (2nd edition) New York: Springer.

Kincaid, D. (2002) *Adapting buildings for changing uses: guidelines for change of use refurbishment*, London: Spon Press.

Kucik, L.M. (2004) Restoring life: the adaptive reuse of a sanatorium, Master's Thesis, University of Cincinnati, OH.

Langston, C. (2007). "Application of the adaptive reuse potential model in Hong Kong: A case study of Lui Seng Chun." *International Journal of Strategic Property Management* **11**: 193-207.

Ohemeng, F. (1996) The application of multi-attribute theory to building rehabilitation versus redevelopment options, in proceedings of COBRA, RICS, Las Vegas, NV.

PCA (2007) *Benchmarks survey of operating costs: Melbourne office buildings*, Melbourne: Property Council of Australia.

PCA (2008) *The office market report*, Melbourne: Property Council of Australia.

Remøy, H. and Van der Voordt, T.J.M. (2006) A new life: transformation of vacant office buildings into housing, in proceedings of CIBW70 Trondheim International Symposium, Norwegian University of Science and Technology, Trondheim, Norway, June 12–14.

Remøy, H. and Van der Voordt, T.J.M. (2007), A new life: conversion of vacant office buildings into housing, *Facilities*, 25(3–4), 88–103.

RP Data (2008) *Melbourne Cityscope*, Milsons Point: Cityscope Publications Pty Ltd.

Swallow, P. (1997) Managing unoccupied buildings and sites, *Structural Survey*, 15(2), 74–79.

Wilkinson, S.J. (2011) The relationship between building adaptation and property attributes, PhD Thesis, Deakin University, Burwood, Australia.

Sustainable Adaptation: A Case Study of the Melbourne CBD

4.1   Introduction

This chapter sets out ten case studies from the City of Melbourne's 1200 Buildings Programme to gain a deeper understanding of the sustainable building adaptations in practice. The context for sustainable building adaptation in Melbourne is described, followed by the typical sustainability measures implemented in office adaptation. Case study research is either exploratory or explanatory (Robson 2003). The primary purpose of the case studies was to observe and describe what measures had been undertaken. External validity issues centre on the representativeness of the cases and how they can be extrapolated to the wider population. In this chapter, all the cases posted on the City of Melbourne 1200 Buildings Programme website were analysed, and the research has external validity because all cases are considered. The analysis is a census of all the projects completed to date within the programme for which data is available. Each case is described with the objectives, challenges and sustainable adaptation measures undertaken and outcomes discussed.

4.2   The Context for Adaptation

A key driver globally for sustainability comes from government. Using Australia as an example, government exists at three levels: federal, state and local. At a federal level the government has set a target of reducing carbon emissions by 5% of the 2000 levels by 2020 (Australian Government Department of Climate Change and Energy Efficiency 2010). The City of Melbourne aims to be carbon neutral by 2020 (Arup 2008), an ambitious

Sustainable Building Adaptation: Innovations in Decision-Making, First Edition.
Sara J. Wilkinson, Hilde Remøy and Craig Langston.
© 2014 John Wiley & Sons, Ltd. Published 2014 by John Wiley & Sons, Ltd.

target in a country where per capita carbon emissions are high at $18.75CO_2e$ t/person/year (Carbon Planet 2011). Melbourne has Australia's highest per capita carbon emissions. The strategy the City of Melbourne has developed comprises a variety of measures such as carbon trading, reductions in transport-related emissions and, after considering the performance of commercial stock, building adaptation. A target has been set of adapting 1200 commercial buildings and incorporating sustainability initiatives to reduce greenhouse gas emissions from the sector (Lorenz et al. 2008). The City of Melbourne is taking an innovative and proactive approach to establish strategies to deliver sustainability in the built environment within the 2020 time frame. It has estimated it is possible to reduce the overall carbon emissions for the CBD by 38% or 383,000 t of CO_2e through building adaptation (Arup 2008). Such a reduction would constitute a significant proportion of overall Victorian emissions, although recent reports reveal an upward trend in emissions (Melbourne City Research 2006; Australian Institute of Urban Studies 2007). In the last 2 years or so, the programme has been affected by a general reduction in construction activity in Melbourne resulting from the global financial crisis (Wilkinson 2012).

Given the upward trend in emissions and aspirations to reduce building-related greenhouse gas emissions, one question is whether the target set by the City of Melbourne for building adaptations is achievable. A snapshot of the Melbourne office market in July 2008 indicated a total of 34 building projects were being undertaken in the CBD; of these 11 were classed as full or partial refurbishments (PCA 2008b). The City of Melbourne envisages that policies and programmes implemented by 2012 will lead to 1200 building adaptations before 2020 (approximately 150 per annum). To achieve this target the adaptation rate has to increase substantially. This, in turn, begs the question: which 1200 buildings? Is it medium-sized, small or large buildings? Other questions arise such as the following: Could the city identify which buildings are most likely to be adapted prior to 2020? More importantly, how do you decide which buildings should be adapted? There is a need to address these knowledge gaps for the city to fulfil its aspirations. The research described in Chapter 3 sought to identify where local policy-makers could use previous practices to best identify which sectors of the property market might optimise sustainable building adaptation. Other cities globally face similar issues and knowledge gaps with respect to building and climate change adaptation.

4.3   Typical Sustainability Measures Used in Commercial Building Adaptation

Using the frameworks outlined in the environmental assessment tools widely adopted around the world, sustainable retrofit measures can be categorised as energy related including emissions, embodied and operational energy consumption, management and auditing, reporting and targeting. With water, issues are conservation and consumption, recycling, harvesting and maintenance of installations. Hazardous materials are covered as are waste and health and well-being issues such as air quality, noise, lighting and thermal control. Pollution and transport-related issues are also included in many rating tools. Finally, land use and ecology issues such as impacts on biodiversity are considered in tools such as BREEAM. This section outlines typical sustainability measures applied to office stock adaptations.

The most important measures are energy related, that is, aimed at reducing consumption and emissions. The measures are further classified into building fabric measures and services measures. Energy measures to the building envelope can comprise the sealing of leaky envelopes, roofs and walls, to prevent heat loss or in the case of hot climates where cooling is required loss of air-conditioned cold air. In a similar vein, insulating the building to reduce the heat/noise transfer can be undertaken. Insulation can be applied externally, in the form of over-cladding, internally through dry lining and within cavities where they exist. Another option available to designers is to configure the space to work with the climate, for example, siting meeting rooms on external walls where temperature fluctuations from heat gain do not affect permanent office users located within the central part of the floor plate. This design approach has been labelled 'virtual double glazing'.

The provision of a green roof or rooftop gardens can also improve thermal performance of the roof to some degree as well as enhancing social sustainability, if building users are allowed access to use the space. Green facades can contribute to thermal performance albeit to a limited extent, though greater benefits are derived from the feelings of well-being experienced by users. Retrofitting external shading devices to elevations is a good option to deflect heat gain from a building, which can take for the form of automated blinds which track the sun's path or a fixed shade. Similarly, fixing shading to roofs will reduce heat gain and decrease demands on cooling systems. Painting roofs white is a relatively inexpensive way of reducing heat gain that is adopted in countries where cooling is a priority.

On-site microgeneration is another means of reducing emissions, and the installation of PV solar supplying a percentage of building power is a way of achieving this outcome. Wind turbines are another means of on-site energy generation. Furthermore, the purchase of 'green power' from energy providers is a popular means of offsetting emissions.

Building services offer great potential for reducing energy use. The options for sustainable adaptation include the provision of gas-fired boilers replacing high-carbon coal-fired installations for heating and cooling installations. Another measure is to fit web-enabled building management systems (BMS) that monitor the building services to ensure optimum energy-efficient operation. For example, the BMS can also work to open and close windows automatically to ensure heat is retained and expelled as temperatures change; this is known as automated night flushing or purging. In some buildings sub-metering of space gives tenants greater control of their energy usage. Where lighting is concerned, contemporary technology is more energy efficient, and the current specification of commercial buildings is for T5 lighting, provision of motion sensors and low-energy lighting (such as fluorescent and LED lighting) in common areas. Furthermore, LED and fluorescent lamps can be connected to intelligent control systems to reduce total consumption.

Cooling office buildings accounts for considerable energy consumption, and retrofit measures include the installation of chilled beams that use water piped through ceiling-mounted pipes where heat is exchanged. The advantage is that air is not piped through air conditioning and occupant health is improved with the chilled beam technology. Other measures around cooling include specifying high-efficiency chillers, variable speed drive (VSD) and air handling units (AHUs) with economy cycles. Variable refrigerant volume (VRV) air-conditioning systems also operate more efficiently than older installations. Zoned floors can also reduce overall energy demand by providing heating and cooling where required. Where mechanical cooling systems are not provided, the provision of ceiling fans can provide a low-energy alternative. Another option is to use new heating, ventilation and air-conditioning (HVAC) systems with high star rating HVAC inverters and with sensor controls.

Where water services are concerned, retrofit options include flow restrictor taps in washrooms and kitchens. Solar power for water heating also reduces emissions. Alternatively heat as required gas water systems result in lower emissions per litre of water heated. Rainwater harvesting systems allow water to be used in flushing toilets and watering plants in a building, thereby reducing consumption of potable water. In addition, waterless urinals and dual-flush cisterns can reduce water consumption.

The reuse and recycling of building materials lowers embodied energy in the building and reduces waste to landfill, as does new construction with recyclable materials. Furthermore, users can adopt recycling programmes to reduce in use waste. Hazardous materials should not be specified such as volatile organic compounds (VOCs). Health and well-being issues relate to air quality such as number of air changes and amount of fresh air, noise levels and acoustic performance, the type of lighting specified and its energy intensity and, finally, the level of thermal control the individual office user has over their space. Pollution measures include control of groundwater, light pollution, land contamination, refrigerant leakage monitoring, control of emissions to air and flood risk management. Transport-related measures include proximity to public transport, provision of shower facilities and bike storage for users and a reduced amount of on-site car parking. These measures are not exhaustive but give an indication of what is currently undertaken globally.

4.4   Sustainable Adaptation Case Studies

Ten case studies of sustainable building adaptation, located in Melbourne, are described. All photos in this section are provided by courtesy of Ruskin Black.

4.4.1   131 Queen Street

Originally built in the early 1900s, 131 Queen Street underwent numerous adaptations including two extensive ones during the 1930–1950s where eight floors were added. In 1955 the facade was rebuilt. The net lettable

Figure 4.1   Front elevation, 131 Queen Street.

area (NLA) comprises 5830 m² of space and accommodates office use, a Buddhist art gallery and cafe, a turf accountants bar and a restaurant. Eleven owners that comprise an owner's corporation may have made decision-making in regard to adaptation more complex and time consuming. The works were undertaken from 2008 to 2011 and predate Mandatory Disclosure legislation that requires owners to obtain a NABERS Energy rating for buildings exceeding 2000 m². Total costs were $1.5 million, and savings of $50,000 per annum were estimated as a result of the energy-saving measures introduced (Figure 4.1).

The objectives were:

1. To make safety and essential services code compliant
2. To upgrade to a 4 to 4.5 NABERS Energy rating
3. To significantly reduce running costs by focusing on preventative maintenance

The sustainable adaptation features were:

1. Sealed roof membrane
2. High-efficiency chiller

3. VSD, AHU
4. Economy cycle
5. Digital BMS
6. Award-winning rooftop garden
7. New fire panel and HVAC system
8. Installation of motion sensors and T5 light fittings and globes in most common areas

Existing water consumption was low and not addressed in this adaptation. Other measures, such as a waste programme, were limited by the owner–tenant structure of the building. Social sustainability was covered in the provision of a rooftop garden for occupiers. The garden mitigates the urban heat island effect, reduces storm water run-off and insulates the upper floor as well as providing social space.

Perceived problems were poor-quality air conditioning that had not been adequately serviced or maintained. This problem was compounded with the need to get all 11 owners to agree to measures and costs; where some had undertaken upgrade independently, they were reluctant initially to commit further funds. Value for money was important. The negotiations took a year to complete and were helped with a $500,000 grant from the Australian Government from the Green Building Fund. It is possible that this project would not have been completed, either to its current standard or, at all, without the financial assistance that was a third of the total costs. Work was completed over 10 months; a requirement of the grant was the work should be completed within 12 months. To minimise disruption to tenants, work was undertaken at night and on weekends which increased costs. Another factor driving up costs was that access to the property for contractors was from the street entrance only.

The outcomes of a 40% reduction in energy costs are predicted; the owners are waiting for a 12-month period to lapse before they can quantify the annual savings accrued and thereby release the final 20% stage payment of the Green Building Grant. Secondly, the green rooftop is much valued and used. Savings of $50,000 per annum are anticipated with the new services requiring less maintenance. The key issue with this case was the complex ownership structure.

4.4.2   Alto Hotel (636 Bourke Street)

This adaptation transformed a heritage building into a low-carbon emission hotel. Designed in a 'neo-Baroque' style, it was built mostly of brick, with granite and bluestone façades and floors of New Zealand kauri pine. The building was substantially renovated and enlarged towards Little Bourke Street in 1981. It was sold to the current owners in 1999 who built a new six-storey structure at the northern end of the building in 2005 and redeveloped the original Bourke Street section. The building is six storeys high with a floor space of 2800 m², with each floor plate being 480 m². High buildings on the eastern, western and northern sides protect the building. The southern

Figure 4.2   Front elevation, 636 Bourke Street.

façade retains original windows. The new building opened in February 2006. Although many of the early architectural features have disappeared, the building is of state significance, and it was placed on the Heritage Register in 2005 (Figure 4.2).

The primary objectives were to develop an environmentally efficient building, in the use of energy and water, and to minimise noise transfer in and around the building. In 2005, NABERS and Green Star were not available, and the owners adopted two other energy and water efficiency measures. An annual EarthCheck audit under the auspices of Green Globe provides a certification programme. The EarthCheck Program was developed by the Australian Government-funded Sustainable Tourism Cooperative Research Centre and is used widely in the tourism industry. Secondly, there is an annual CO_2 audit under the auspices of the Carbon Reduction Institute. This was established in 2008 for the purpose of promoting awareness and action on climate change and provides a 'No CO_2' certification programme for member organisations and is used widely in the tourism industry. The sustainable adaptation measures were:

1. Insulating the building to substantially reduce the heat/noise transfer.
2. High star rating HVAC inverters, with sensor controls.

3. Heat as required gas water.
4. Hot-water reticulation system.
5. Low flow taps and showers.
6. Fluorescent or LED lamps.
7. Refillable dispensers in hotel rooms for guest complimentary toiletries.
8. Organic waste and frying oil disposal for conversion to biodiesel fuel.

Given that each room has its own independent inverter system for heating and cooling, there is no need for a central BMS. The hotel has a 100% green energy acquisition policy and is seeking to become zero carbon. With this adaptation, the innovative construction techniques needed to provide substantial heat and acoustic insulation that was unfamiliar initially to the building contractors were the biggest challenge.

The outcomes were that:

1. Energy consumption per guest was reduced to 36 MJ/day, 78% better than best hotel practice.
2. Water consumption is 123 L/day per guest (68% better than best hotel practice) and the financial savings from reduced water use and water heating costs are significant.
3. The sustainable attributes of the hotel make it attractive to guests scoring well in terms of social sustainability.
4. Maintenance costs may be higher than previously.
5. The owners believe that energy efficiency of the building fabric is paramount and should be addressed before the services installations are changed.

Further work is planned to use solar power to heat the commercial kitchen's hot-water systems. There is insufficient roof space to provide hot water for the whole building.

4.4.3   247 Flinders Lane (Ross House)

Built in 1897 as a six-storey warehouse and heritage listed, Ross House has an NLA of 2120 m². The building consists of a basement car park, a small retail shop on the ground floor, five office floor levels and a roof plant room. The building was designed as a brick structure using a Romanesque style, with brick arcades, metal oriel windows and parapet colonnade. The building was converted to offices in 1931. In the mid-1930s, the Flinders Street half of the building was demolished and new offices built. The Flinders Lane building was retained and named Royston House. Ross House, as it was renamed, opened in 1987. The building was purchased through a grant from the R.E. Ross Trust by the Ross House Association, which offers low rent to tenants; mostly small, independent community and self-help organisations committed to social justice and environmental sustainability.

In the last adaptation (1985–1987), all systems were upgraded. The HVAC was changed from a central system to individual units. An environmental management plan (EMP) identifies and implements adaptation works over time as funds become available and the adaptation may run over 4 years due to funding issues. In addition, heritage issues have to be considered, and the Conservation Management Plan covers these aspects. The 'low-hanging fruit'; that is measures mostly easily and cost effective have been implemented from the EMP.

One issue was airtightness as the 1901 building fabric was leaky. The HVAC had reached the end of its life cycle and needed replacing; however, funds were not available. Switches had to be positioned so that each tenants' energy load could be established. Water consumption patterns were already very efficient, and due to financial restrictions and prioritisation of other measures, no further water efficiency measures were required. Waste separation is encouraged in the building with separate bins for recycling waste. The work undertaken by the tenants of Ross House meant that there is a high degree of social sustainability associated with the property. The objectives were to overhaul all systems and to bring the building to a minimum NABERS 4-star level. The key challenge with Ross House was financing the adaptation and the association sought grant funding for the $500,000 required. The associated challenge was then to provide the optimum environmental benefits at an affordable cost (Figure 4.3).

Figure 4.3   Front elevation, 247 Flinders Lane.

In this project the sustainable adaptation features were:

1. Installation of time clocks for each instantaneous boiling unit.
2. De-lamping about 50 existing T8 twin light fittings.
3. Installation of light sensors.

To date no outcomes for Ross House have been reported as the adaptation was planned for 2012.

4.4.4   490 Spencer Street

This example shows the efficiency gains made by reducing the heat load in an older small office building prior to upgrading the mechanical systems. The property is a two-storey office at the north end of Spencer Street. It was built in the 1980s comprising tilt-slab construction with no insulation and dominated by large west-facing glass windows and a grey rendered concrete façade. The HVAC system was a reverse cycle centralised package, with two units cooling and heating the two floors separately. Tenant comfort was achieved by using large amounts of energy through the HVAC.

The decision-making for the adaptation comprised three stages. The first step was to undertake a thorough investigation to understand any building performance problems. The unique attributes of the building were identified, particularly good aspects, and showed the back and base of the building were surrounded by concrete, producing a 'cave' which had potential. The limitations were many ventilation leaks and no insulation. Stage two determined how much energy could be retrieved from zero-cost sources. Modelling indicated that the owners should be able to obtain 50% of energy needs from other sources: using outside air as an air-conditioning economiser and air trapped in the ceiling cavity for heating in winter. Systematic lowering of energy use was continued by introducing low-energy computers and reducing the number switched on at any one time, more efficient switching, providing an individual workstation power board, so all power is connected and one switch turns it off. The aim was to make 'no regrets' changes justified on economic grounds alone. The owners wanted a 'high-tech' building because they believed it removed the chance for occupants to interface with controls resulting in lost ownership over energy-saving efforts. Occupant engagement was critical in achieving the educational benefits of green buildings. Stage three was to reduce energy demand and install an HVAV system to replace the existing one (Figure 4.4).

The adaptation commenced in mid-2009 with the tenants in occupation. The owner believes a combination of a practical architect, a focus on engineering and aesthetics and highly qualified tradespeople was vital to the success of the project. Effective cross-disciplinary communication was also critical. A survey showed where the leaks were and they were plugged. R3 insulation was retrofitted above suspended ceiling tiles, as the cost of replacing the roof (with the air-conditioning plant thereon) was prohibitively expensive.

Figure 4.4   Front elevation, 490 Spencer Street.

The 'virtual' double glazing comprises the installation of rooms for short meetings on the western elevation with glazed partitions to allow light to penetrate into the office spaces. This side of the building was subjected to solar gain that was cooled by the air conditioning. Virtual double glazing means the short-term meeting rooms collect the heat otherwise distributed into other office areas and reduces the air-conditioning load. This was a low-cost solution. New external blinds on the western elevation further reduce the internal heat load. The HVAC requires replacing as the energy load has been reduced a smaller system will suffice, however there are insufficient funds to include this measure in the current adaptation. The adaptation reduced energy loads through the provision of energy-efficient lighting and appliances, PV cells on the roof forming a 10-kW PV array, which provides 20% of the original building power needs and up to 100% of the adapted building power needs. Other measures included energy-efficient desktop computers with low-energy hard drives and using laptops, that consume a quarter of the power, in preference to PCs, remanufactured low-power Fuji Xerox printers, second-hand fridges and high-efficiency dishwashers. Furthermore lights out stickers, lighting timers, zoned lighting in the open plan office, motion sensors, T5 fluorescent lights in office spaces, LED in foyer and conference rooms, large windows to maximise natural light, some new double-glazed windows, a skylight was added to upstairs office to allow natural light and 100% green energy all contributed to reductions in energy consumption.

Water economy measures include WCs with a 60-L flush and cisterns filled from washbasins. Only cold water is fitted to the kitchen and bathroom sinks to avoid energy wastage by heating hand water.

Waste was reduced by purchasing second-hand office furniture and carpet where possible, and reusing materials from pre-existing offices, other furniture was made from existing components, and finally paper used is 80% recycled and Australian. In the adaptation 'redundant' materials were left in the laneway behind the building – all were taken and there is a scope for a city-wide scheme. Wider efforts were made to change behaviours through the provision of bike racks and showers for occupants. Indoor plants were provided to enhance indoor air quality and serve as natural visual and sound barriers.

The objective was to create a zero greenhouse gas emission building. Energy efficiency objectives can be compromised by other concerns such as occupant behaviour. In this adaptation, the owners wanted to avoid these compromises and make energy efficiency the primary focus.

Sustainable adaptation features included:

1. 'Virtual double glazing'.
2. PV solar supplying 20–100% of building power depending on the amount of sunlight and energy use.
3. Energy-efficient lighting.
4. 100% green energy.
5. Water-efficient appliances.
6. Reuse and recycling of building materials.
7. Web-enabled BMS.

The key challenge was the building's inherent poor energy efficiency. Secondly, the cost of installing PV was high, though it should become more viable as economies of scale are achieved over time. The outcomes were:

1. On a sunny day this building is zero carbon.
2. Water consumption is being tracked for comparison to similar stock.
3. Financial savings are considerable with improved maintenance management.
4. Higher rents have been achieved post-adaptation, along with lower operational costs.
5. While challenging, the benefits are profitable and satisfying according to the owners.

4.4.5   500 Collins Street

500 Collins Street is an extensive adaptation to achieve energy and water efficiencies while maintaining high tenancy levels during the works. Completed in the 1970s, 500 Collins Street was renowned for its quality of construction, modern building standards and services, a consequence of which was a high tenancy profile. By 2002, the building had declined to a low B-grade standard through obsolescence and ageing. Despite this decline, the tenants remained

because of building size and configuration, excellent location and sound building management. These attributes led the owner to determine that it was suitable for adaptation. Before adaptation, the building comprised 23,500 m² of office space, five retail shops and 140 car parking spaces. The project began in mid-2003 and was completed in early 2011. The project was delivered in three phases to allow for the almost fully occupied building to operate during adaptation. Phase one was to replace and upgrade plant and to renovate the façade. The second phase was to maximise retail space and reconfigure the car park, while phase three comprised office floor upgrades.

Adaptation was rolled out progressively as leases expired. The work included strip out of each floor and replacement with new finishes and chilled beam air conditioning. The floor-by-floor upgrade required much planning and could only be realised when tenants vacated. Generally, it was achieved three floors at a time. It also meant, as much as possible, minimising the impact of the building construction on the tenants that was accomplished by having several lifts set aside for the builders' use. Demolition was completed out of hours, and old carpets were laid on the concrete floors to reduce noise for occupiers.

A chilled beam solution was chosen for the HVAC. The façade was upgraded by replacing glazed spandrel panels with aluminium wall panelling, repairing and refurbishing vertical columns and repainting. Changes in the floor structure increased the NLA to approximately 24,400 m² along with eight additional shops; a small reduction in car spaces; the addition of bicycle racks, change rooms and shower facilities; and the provision of disabled access and amenities. Foyer entries and public areas were upgraded. Level 3 was extended onto the podium of level 2 to create external meeting space and recreational landscaped areas.

The HVAC systems included the installation of new energy-efficient chillers with VSDs and more water-efficient cooling towers and new gas-fired boilers for heating replacing oil-fired ones. As each floor became vacant, chilled beam air conditioning was installed. This is a combined system with active chilled beams using fans to diffuse cool air around the building's perimeter where solar loads are high and passive in the interior spaces. The original central ducting was reused in the perimeter zones. The new system decreased the number of fans for air conditioning, reducing energy usage. As the building was tenanted, the old system was maintained alongside the installation of the new. Energy load was reduced by installing roof-mounted solar panels for 25% of domestic hot-water supply, fitting low-energy T5 light fittings in all public and leased spaces, installing VSDs on major plant and equipment and using chilled beam air conditioning (Figure 4.5).

Water consumption was reduced by fitting waterless urinals, 3- and 6-L dual-flush cisterns, flow restricting devices on all fixtures, rainwater and condensate capture for landscape irrigation using large tanks in the basement parking area to store water and lastly baffles on the cooling tower preventing aerosol spray. Waste was tackled through on-site recycling bins, with around 80% of construction waste that was recycled. General sustainability improvements included minimising embodied energy of the building; using PVC-free materials where possible; using low-VOC

Figure 4.5   Front elevation, 500 Collins Street.

materials, a preference for materials with a high-recycled content; selecting materials for durability and from sustainable sources; encouraging the use of bicycles by providing a secure bike area for 82 bicycles, plus shower and change room facilities; and improving the indoor environment quality by increasing fresh air by 50%, radiant cooling (chilled beams), low-VOC materials and reduction in indoor ambient noise levels.

The building control system was renewed, with commissioning an ongoing process as each floor was completed. The main electrical switchboard was replaced, and tenancy sub-metering supplied to enable effective energy monitoring. The commissioning of plant and equipment was critical so that building management understood how the building functioned and how to control it for efficient operation.

The objectives were to achieve an A-grade building standard; to attain a high degree of environmental efficiency, both during the upgrade works and post-upgrade operations (set before NABERS and *Green Star* ratings were in place); to maximise tenant retention during the upgrade to maintain optimum cash flow and provide a potential pool of long-term tenants; and, finally, to elevate tenancy profile by increasing the average size of tenancy, length of tenure and quality of tenant to achieve a commercially justifiable return on investment. The sustainable adaptation features were:

1. Energy-efficient VSD chillers.
2. Gas-fired boilers.
3. Chilled beams (passive and active).

4. Solar panels servicing 25% hot-water requirements.
5. T5 light fittings.
6. Water tanks collecting rainwater and condensate for landscape irrigation.
7. Waterless urinals and dual-flush cisterns.
8. Flow restricting devices on all fixtures.

The key challenge was adaptation with tenants in situ. Furthermore, the approach necessitated the continued operation of existing services, while new installations were being fitted. The outcomes were as follows:

1. Energy was modelled to achieve a 30% decrease in air conditioning, a 50% decrease in lighting and a 15% fall in hot-water usage.
2. Water modelled to attain 40–50% savings.
3. Sustainability Victoria and the owner undertook a productivity study in 2007–2008 which observed a 39% reduction in average sick days monthly per employee, a 44% fall in the average cost of sick leave, a 9% improvement on average typing speeds and considerable accuracy improvement of secretarial staff, a 7% increase in billings ratio despite a decline in average monthly hours worked, a 7–20% decrease in head-aches, a 21–24% decline in colds and flu and a 16–26% fall in fatigue; it is held the results are due to improved building amenity and air quality.
4. Reduced maintenance costs are due to reductions in plant and equipment, more efficient plant and improved monitoring through the BMS.
5. The rental value of the adapted space has increased substantially.
6. Over the project, the team maintained support of tenants with occupancy rate not less than 70%; the building now has fewer tenants, meaning the number of larger tenants has increased.
7. The building has a 5-star *Green Star* Office Design v1 rating.
8. The main lesson learned was the importance of tenant communication; secondly, strong project management leadership, so that the team understands the sustainability objectives and works to assess all elements against the ESD criteria; thirdly, meticulous management and control of noise and temporary service shutdowns; fourthly, engaging an ESD consultant, who advocates for the ESD in the project team; and, finally, appoint an independent commissioning agent to specify commissioning and tuning criteria and timing of the project.

4.4.6   406 Collins Street

This project provides a demonstration of how to replace the HVAC in a 1960s midsize office building while maintaining tenancy. 406 Collins Street is a modernist building, typical of the late 1950s skyscraper design, built from a steel and concrete structure, a plain (non-ornamented) façade, with a strip of windows on each floor facing Collins Street. In 1961 the building was extended with four storeys added. The only feature from the original 1897 building is the 'Atlas' statue, a decorative pediment at the top of the building, which is now sited at street level by the entrance. The

Figure 4.6   Front elevation, 406 Collins Street.

HVAC system was typical of the 1960s: minimal capital cost but no concern for energy efficiency. A new owner purchased the property in 2006 and decided it needed substantial adaptation.

The building has 4000 m² NLA, including a retail store on the ground floor, and eight tenants occupying at least one floor apiece. The floor plate is rectangular, measuring about 350 m² to each floor. The HVAC system had reached its use-by date. Modification of the existing system was not considered to be a viable option to achieve a significant improvement on energy efficiency (Figure 4.6).

To improve energy efficiency of the building, to achieve a minimum 4.0-star NABERS Energy rating, to reduce the carbon footprint and to use green power sources were the objectives with 406 Collins Street. Sustainable adaptation features were:

1. VRV air-conditioning system.
2. Zoned floors.
3. Economy cycle dampers.
4. Automated night flushing.
5. Roof sunshade.
6. Internal and external shading in courtyard.
7. Motion light sensors in stairwells and lifts.

8. High-efficiency lighting in common areas.
9. Sub-metering system.
10. Web-enabled building management control system (BMCS).

The main challenges were dealing with the difficulties typically encountered in old buildings and the need to maintain the existing services in a fully tenanted building during adaptation. Another challenge, according to the owner, was managing cash flow during the project. Initially, he tried to fund the project from the ongoing rent, but it did not meet expenses. The Green Building Fund grant provided $500,000 for the whole project, but only 20% or $100,000 up front with the balance on completion. To make up the shortfall, a loan was secured from the Sustainable Melbourne Fund (SMF). To meet the costs, this project could not have proceeded without the building being occupied, and so the challenge was to minimise the impact of the works on tenants. The outcomes were:

1. Energy performance should be halved or as low as 25% prior to the adaptation.
2. The building has low water usage, achieving a 5.0 NABERS water rating.
3. With the new system there will be perceptible variations in the internal temperature range; educating tenants to accept slightly warmer ambient temperatures in summer and cooler in winter will allow significant energy savings.
4. With HVAC improvements made and the installation of the BMCS, the engineers and building management are confident that maintenance will be faster and less expensive.
5. Although it is highly likely there will be significant energy improvements, the owner is unsure that there will be direct financial returns on investment; the viability of the project hinged on the Green Building Fund grant, and without it, he doubts that the project would have been as extensive, and it would have meant an even lower return on investment.

4.4.7   182 Capel Street

A small commercial office building constructed in the mid-1980s, 182 Capel Street is a precast concrete structure supporting a lightweight steel frame supporting concrete flooring. It has two floors and a basement car park. The building was purchased in 2003 with the intention of sustainable adaptation. The net tenantable space is 1200 m^2, but a new second floor will bring this to 1600 m^2. The building faces west, offering excellent exposure to light, but heat on the façade was a problem that needed addressing during adaptation. It is positioned near parkland and combines commercial, retail and residential zoning.

The mechanical systems were at end of their life cycle. Originally, the building had its own generator and substation, but the generator had been removed. Notably around 60% of commercial buildings in the fringes of Melbourne are similar in size and construction to 182 Capel Street.

Figure 4.7   Front elevation, 182 Capel Street.

The project did not follow a set planning path as the owners are familiar with design and construction and paced the work around their business commitments. However, when they received a grant from the Australian Government's Green Building Fund, the project was required to meet deadlines. While the grant does not cover all the adaptation costs, it makes an important contribution. To qualify for the grant, the owner has to adapt the building to reduce carbon emissions from a 1.5-star NABERS Energy rating; the goal is 5 stars.

One aspect of the adaptation is the operable windows that provide natural ventilation throughout. The windows are located on the western and southern elevations and are linked to a weather station that informs the mechanical system to activate the air conditioning on or off and open or close the windows. A cross-ventilation system allows fresh air in during the day and purges during the night. The building was well insulated, so no additional work was required (Figure 4.7).

The main objective was to significantly reduce the building's carbon footprint. The aim is to reduce carbon emissions by at least 50% and attain a 4.5-star NABERS Energy rating. The adaptation features were as follows:

1.   Automated opening windows connected to economy cycle and control system.
2.   Automated external blinds.
3.   Gas-fired VRF gas heat pump air conditioning.
4.   LED and fluoro lamps connected to intelligent control system.
5.   Rainwater collected in (stage II) tanks located in basement for WC flush and irrigation.

6. Green wall (vegetated façade).
7. Bokashi buckets for waste disposal, green façade nutrient.
8. Intelligent component control systems.
9. Additional building sealing and insulation.
10. Ceiling fans.
11. Fit-out and construction with recyclable materials.

The main challenge was deciding what to do initially, after which there were cost constraints to consider. Some ideas that sounded good in theory were critically assessed and rejected because either the costs could not be justified or it was unclear whether they would work. Another challenge was managing staff expectations as to when construction would be completed. There were times when the building work was invasive. Doing the mechanical work and replacing the windows was complicated especially working around the occupants. Detailed programming was required to achieve this efficiently. The outcomes were:

1. A NABERS Energy rating was due to be conducted towards the end of 2011 with a target of 4.5–5-star NABERS Energy (a 50% reduction in carbon emissions).
2. Water economy exceeded 50% reduction, saving 900 L per day.
3. Socially, tenants are happy with the potential for the building and have decided to stay.
4. Mechanical maintenance reduced from $3200 to less than $1000 per annum.
5. Lighting maintenance reduced from $1200 to $500 per annum.
6. Commercially, the project needed to achieve an 8% yield on investment that has been achieved thus far.

The key lessons learnt were to start with the building fabric and see what can be achieved through modifications. Secondly, work within the constraints of a realistic budget to ensure there are real returns on investment. Thirdly, confirm there is a project champion. A fourth lesson was to organise high-level input, contributing ideas before a budget is committed. This means letting the ideas drive the process and allowing the building to be what it should be. Sometimes designers push a building too hard, when it's just not going to work, and money will be wasted. Decisions need to be value managed. Finally, it is necessary to work with the inherent attributes of the building and consider ideas as a team comprising engineers, architects, building managers and users in an 'ESD workshop'.

4.4.8   115 Batman Street

This project exemplifies a building that uses various energy-efficient techniques and technologies to achieve an excellent NABERS Energy rating. 115 Batman Street was originally a machinery factory built in the 1920s and had been derelict since the late 1980s. The factory was gutted, leaving only the

Figure 4.8   Front elevation, 115 Batman Street.

brick elevation. The new building was constructed in this space, combining new and existing. The basement car park and two of the floors are within the original elevations, while two new floors were added. The new area above the original brick walls comprises Vitrapanel clad with well-insulated walls and a new insulated pitched, profiled metal roof. The two lower floors utilise the original space in the brick elevation for the windows. The free-standing building faces north and has very little shade provided by surrounding buildings (Figure 4.8).

In summary the objectives were to introduce state of the art engineering services with very low-energy consumption, to provide a comfortable working environment to enhance productivity and to achieve 5-star *Green Star* and 5.0-star NABERS Energy ratings. The sustainable adaptation features were:

1. Complete reconstruction within existing façade.
2. Highly insulated building shell.
3. Chilled beams in the ground, first and second levels.
4. VAV economy cycle on the third level.
5. High-efficiency gas boiler for heating.
6. High-efficiency luminaries.
7. 15,000-L rainwater tank.
8. Solar panels for water heating.

The project outcomes were:

1. The chilled beams coped well with the March 2009 heat wave in Melbourne. They responded well to the changes in ambient conditions and were superior to the third floor VAV system.
2. The lighting of base building consumes less than 2 w/m^2/100 lux.

3. The building exceeds 5.0-star NABERS Energy performance at 89–91 kg/m²/year. The NABERS rating benchmark for 5 stars is 101 kg/m²/year.
4. Socially, there is very positive feedback from the staff about the workplace environment.
5. The system is simple, the plant is accessible, and the maintenance is straightforward.
6. Total building outgoings are less than $60/m² which compares with most office buildings in the city where the outgoings range from $70 to $90/m².

4.4.9   385 Bourke Street

385 Bourke Street is an example of how to manage the adaptation of a very large office building over a long project period to achieve significant improvements in energy efficiency. The building has a concrete and steel structure with rectangular windows in a concrete façade. Completed in 1983, it is 45 storeys high, sited at a 45-degree angle to the road. The building houses many commercial tenants, with retail space and a large food court on the lower levels. The retail area is 6000 m² and the office is 55,000 m².

In 2004, an ABGR rating (the predecessor to NABERS) revealed a 0-star rating. An environmental performance audit provided a list of opportunities to improve energy efficiency. Some of the mechanical systems had reached the end of their life cycle; the recommendations were mainly directed at upgrades to the efficiency of the HVAC (Figure 4.9).

The key objective was to lift the building from a 0 NABERS Energy rating to a 2.5 rating. This objective was essential for the building to maintain relevance in the marketplace, and the sustainable adaptation features were:

1. Upgraded BMCS.
2. Variable speed fan drives.
3. Economy mode.
4. Lux metre sensors.
5. T5 lamps.
6. Quantum heat pump units.
7. Flow restrictors in washrooms.
8. Commingled recycling programme.
9. Metering.

Effective communication in the project team between consultants and contractors was challenging at times. Another concern was getting the installed equipment and control strategies working and tuned correctly post-commissioning and an external technical agent assisted with tuning the building performance. Performance-based contracts were not used due to the amount of additional work going on simultaneously. It would have been difficult to separate the contribution individual component projects were having on the whole energy performance unless all work was rolled under the same contract.

Figure 4.9   Front elevation, 385 Bourke Street.

Commissioning took over a year as there were problems getting the BMS running effectively. There were complex control strategies that required seasonal tuning. Furthermore, parts of the installation works had technical issues relating to latent conditions, and there were differences in perception of scope.

Another challenge was the site documentation was not current, and this was a problem because it was unknown whether the new control strategy would impact negatively on the operation of a different component in the complex system. The project team decided to write the documented programming specification and revised the building technical manual. This made clear what changes were required to all elements of the programme and helped in having accurate as-built manuals.

Maintaining occupancy rates was a challenge. This relates to the NABERS Energy rating as it is calculated on the energy used, square metre occupied and the hours of operation, which means, to gain a significant rating, the building needs to be substantially occupied. The outcomes were:

1. 372 MJ/m^2 annual saving, a 41% decrease in CO^2 and a NABERS Energy improvement from 0 to 3.5 stars.
2. NABERS water rating of 3.5 stars.

3. The works brought up maintenance issues that are being addressed, and management needs time for the maintenance works to stabilise in order to determine the cost impacts and determine a decrease in HVAC comfort complaints.
4. The increase in NABERS Energy rating opens up the building to a larger market of tenants, especially those looking for a NABERS rated tenancy.

4.4.10   530 Collins Street

530 Collins Street was vacant and adapted to meet contemporary standards, including energy efficiency. The owner, GPT Wholesale Office Fund, reported 92% of the tenants rated sustainability as 'very important' or 'important' to their business. The development took a broad view of sustainability making social and environmental improvements. As such improvements included coffee shops, informal meeting spaces and increased food options. 530 Collins Street was built in 1989 with an NLA of 65,775 m². The pre-adaptation NABERS Energy rating was 4 stars, and this was achieved using 25% green power. The NABERS water rating was 3 stars (Figure 4.10).

Figure 4.10   Front elevation, 530 Collins Street.

The objective was to achieve a NABERS Energy rating of at least 4.5 stars and reduce annual greenhouse emissions by 40% compared to the industry average.

4.5   Comparative Analysis of Sustainable Adaptation Measures

This section summarises the sustainability measures implemented in the adaptations. Seventy measures were implemented across the ten cases. They can be categorised as environmental and social sustainability measures, though many environmental measures were executed due to potential economic benefits. Not surprisingly, 61% of measures related to the building services. In all, 73% of measures were energy efficiency related, not only reflecting the importance of energy efficiency in sustainability especially in the fulfilment of NABERS Energy and *Green Star* ratings, but also revealing the poor energy performance of existing stock.

Water economy measures featured eight times in the adaptations. Possibly water economy is not as important as energy or, more likely, that due to water restrictions imposed during the 10-year drought in the early 2000s, many Melbourne buildings operate efficiently in terms of water. A number of cases noted existing water economy was good.

Measures to the building envelope featured 12 times and are associated with energy efficiency. Opportunities for building envelope measures occur less often, involve access challenges and disruption to occupants and are expensive. However, once undertaken, these measures are a more long-term solution than upgraded services which require maintenance and will be replaced typically within a 20-year life cycle.

Social sustainability was mentioned in four cases mostly in respect of amenities provided to users in respect of improved internal environmental quality (IEQ). One project featured a rooftop garden that provided a pleasant social space; however, the rationale for inclusion included environmental benefits of reducing the heat island effect, insulating the roof and reducing energy use (also an economic benefit). Finally, one project featured a building that housed small businesses that were driven by social justice and equity issues, thereby having a positive social sustainability contribution. Overall social sustainability has a lower profile within these adaptations, which is understandable given the main goal of the 1200 Buildings Programme is to reduce carbon emissions. Table 4.1 summarises the cases and whether they had financial assistance from the Green Building Fund, adopted out of hours working and their social, economic and environmental features. Table 4.1 shows there may be more potential for participation in the Green Building Fund and that environmental and social sustainability featured strongly.

Having described the sustainable adaptations, this part of the chapter examines the similarities and differences between cases. Using criteria previous studies have found to be important, this section analyses adaptation attributes and compares and contrasts their relative importance.

Table 4.1 Case study summaries.

Case study	Assistance of Green Building Fund	Out of hours working	Economic features	Environmental features	Social sustainability
131 Queen St	X	X	X	X	X
636 Bourke St	—	—	—	X	X
247 Flinders Ln	—	—	—	X	X
490 Spencer St	—	—	X	X	—
500 Collins St	—	X	—	X	—
406 Collins St	X	X	—	X	—
182 Capel St	X	—	—	X	X
115 Batman St	—	—	X	X	X
385 Bourke St	—	—	—	X	—
530 Collins St	—	—	—	X	—

4.5.1   Owners

Swallow (1997) found owners were important in adaptation with different drivers to act. The cases confirm there were different owner types who acted with different priorities. Five of the ten projects had owners or tenants who were directly involved in the construction industry with a specific interest in sustainability. Furthermore, they used the adaptation as an opportunity to develop their knowledge and expertise in sustainable adaptation to market to clients post-adaptation. In these cases there is self-interest motivating actions as well as a commitment to sustainability. Of the remaining five, two were committed to environmental and social equity issues through their work. For these parties, sustainable adaptation offered the opportunity to 'walk the talk' and demonstrate their commitment tangibly. Three owners were institutional investors, traditionally interested in economic performance of assets within their portfolios. For these parties, the motivation was to maintain and increase the profile and attractiveness of the asset in the marketplace, so the drivers can be said to be economic, environmental and social.

4.5.2   Age

As buildings age, they become worn and subject to obsolescence (Douglas 2006). Without adaptation, buildings affected by obsolescence attract fewer tenants, lower rental income, eventually becoming unlettable and requiring demolition (Swallow 1997). Many cited the commercial case for adaptation as a driver. Wilkinson (2011) found the relationship between adaptation and building age was strong and the primary consideration for stakeholders involved in building adaptation is the age of the building and services.

Half the adaptations occurred to stock aged 19–41 years of age, indicating the buildings were undergoing the first major adaptation. Buildings require some extensive remodelling after their first 20 years, and the fabric and also the space plan age to a point where reconfiguration and renewal is necessary (Brand 1994). The mean age of the 1200 Buildings Programme cases was 43.9 years, which is older than the mean for the entire commercial stock for Melbourne of 31 years (Jones Lang LaSalle 2008), and it is logical that these older buildings are undergoing adaptation. Ten per cent of adaptations occurred to buildings aged up to 18 years, and this indicates that the newer stock suits the needs of contemporary users much better than the stock aged 19 years plus. The literature typically refers to the first major adaptation being required at 25 years or so (Brand 1994; Douglas 2006), and it was found that Melbourne buildings aged over 42 years accounted for 40% of adaptations. The majority of buildings eventually reach an age in Melbourne's market where they no longer meet the market demands even with adaptation. These findings show owners and designers should be realistic of a building's life cycle and avoid over-specification of commercial office buildings whereby extra resources are committed to buildings to provide a hypothetical life of 100 plus years. In reality societal tastes, needs, perceptions and expectations will have changed so much so that the building will be perceived to no longer meet market expectations after this time. In terms of sustainability, this means that consideration of building 'deconstruction' must be a higher priority. 'Deconstruction' is the ability to partially or wholly dismantle or 'deconstruct' a building or parts thereof, during or at the end of the useful life cycle, and it allows for the reuse of building components and/or recycling of those materials for further use elsewhere. In essence the building's life cycle of components is extended for use in other buildings or structures. It can be argued that adoption of deconstruction is a move away from the philosophy of adaptation; however, the argument is that the concept of adaptation as being within building and within site is extended to embrace across building and across site applications; thereby the whole life cycle of materials and components is fully utilised. Adaptation will be more easily accommodated if deconstruction has been considered in the initial design.

Compared to Wilkinson's earlier study (2011) where 71% of adaptation works occurred to buildings aged 19–41 years old, this figure is higher than the 1200 Buildings Programme case buildings. Where younger stock is concerned, the percentage of works is similar; 7% of adaptation works occurred to buildings aged up to 18 years. The major difference is in the older stock where 23% in the first study were aged over 42 years. The 1200 Buildings Programme attracts older stock, though with such a small sample, this cannot be statistically supported.

4.5.3   Location

The Melbourne CBD is divided into five locations (prime, low prime, high secondary, low secondary and fringe). Rents and capital values per metre squared typically decline as one moves from better to less good locations. In work conducted and reported by Kincaid (2002), Douglas (2006) and Highfield (2000), location was important. Five cases were located in the low-prime areas, with one case in the low secondary and four in the fringe areas. Low prime is the second ranked location. The finding compliments

Wilkinson's (2011) study analysing 7393 CBD adaptations from 1998 to 2008 where adaptations were more likely to occur in better locations. In that study 26% of adaptations were reported in low-prime locations, and here the percentage is much higher (50%), though the sample is small. A difference occurs with the low secondary adaptation rates; in the earlier study it was the highest ranked location at 27%, while in sustainable adaptations it ranks the lowest at 10%. With the fringe activity in the second study, the reverse is true; the rate of adaptations is 40% of the total compared to 9% and possibly reflects a drive to enhance the stock in the location and possibly the owner and tenant composition of the latter study.

4.5.4   Aesthetics

Aesthetics was important in adaptation (Chudley 1981) and assessed on the basis of massing, form, composition, use of materials and so on (Zunde 1989). Six of the sustainable adaptation cases were very aesthetically pleasing; one was classed as neutral and three cases were deemed 'unattractive'. This finding aligns closely with Wilkinson's (2011) study where 63% of stock ranked aesthetically pleasing was adapted. At the other end of the scale, in the previous study 17% of unattractive buildings were adapted, and here a higher proportion of 30% was adapted. In most cases little work was undertaken to change the buildings' appearance externally. Again this is likely to be due to the owner/tenant motivation of the case study group. Overall some findings are consistent with Wilkinson (2011) and Ohemeng (1996) in terms of aesthetic qualities of adapted stock.

4.5.5   Location of Vertical Services

Gann and Barlow (1996), Snyder (2005) and Szarejko and Trocka-Leszczynska (2007) found that the location of the vertical services was significant in adaptation. Here most buildings had services located centrally (four) and also to one side (four) within the building, followed by multiple locations (two). This reflects the scale and age of the buildings in the sample. Compared to Wilkinson's (2011) study, most adaptations occurred to stock with services in central locations (56%), followed by multiple locations (34%). This study shows an increase in the prevalence of stock adapted where the vertical services are located to one side of the building. There is some consistency with the highest rate of 40% being to buildings with centrally located stock. Again the sample size is small, and results should be treated with caution.

4.5.6   Existing Land Use

Half the adaptations occurred to buildings classed as sole office land use; 30% for office and retail land use; 10% to office, retail and residential; and 10% to hotel use. This is consistent with Wilkinson's (2011) study where

53% of all Melbourne adaptations occurred to 'office' only buildings. When other land uses are examined, the profile changes; in the earlier study 45% were office and retail land use, higher than the sustainable adaptation study. Furthermore, in the earlier study 2% were attributed to other land uses, whereas here it is 10% and then 10% to hotel land use. A conclusion is a wider range of land uses are drawn into sustainable adaptation with the 1200 Buildings Programme.

4.5.7   Floor Area

Typical floor area relates to physical dimensions of buildings or size (Kincaid 2002; Arge 2005). Floor sizes were divided into small, medium, large and extra-large categories. Sixty per cent of all works occurred to buildings with a typical floor area of 700 m^2 or less (small), with 20% to buildings with a medium typical floor area of 701–1178 m^2. Large typical floor area adaptations (1179 m^2 plus) accounted for 10% of works, with extra-large typical floor areas of over 1346 m^2 explaining 10% of works. Floor size is important in adaptation with floors that are able to accommodate user needs and market demands most likely to be adapted. Wilkinson (2011) found that owners and occupiers adapted floors (and buildings) to suit changed needs regardless of size and that the Melbourne market had a more or less equal demand for all groups of floor sizes. In that study the percentages were 24% (small), 27% (medium), 20% (large) and 28% (extra-large) of works. The sustainable adaptation cases have a different profile reflecting smaller-scale projects that may be a reflection of access to funds and the type of work undertaken post-2008.

4.5.8   Street Frontage

Street frontage measures building width in metres. Building width is an important criterion in adaptation; Povall and Eley's study (cited in Markus 1979) established a benchmark for building width in adaptations. In the Melbourne stock, buildings can have over 200 m of street frontage. In this study 20% were less than 10 m wide, 10% 11–20 m, 10% 21–30 m, 20% 31–40 m and finally 20% 61–70 m. Overall the sustainable adaptation cases comprised building widths mostly in the lower range – up to 50 m wide as in the earlier Wilkinson (2011) study. No very wide buildings featured.

4.5.9   Historic Listing

Historic listing has an important effect on adaptation. Due of the date of settlement of Melbourne and the age of the buildings, it follows that this is an area where a high amount of buildings have heritage overlay or listed status. In the 1200 Buildings Programme study, only 20% of adapted buildings had historic listing or heritage overlay which contrasts to the 2011 study (Wilkinson 2011) where heritage overlay/listed stock experienced higher rates of adaptation than the non-listed stock. Where overlays exist or listing occurs, owners are required to heed requirements and obligations established by legislation in respect of the properties' external appearance or materials used and/or also with regard to building interiors in some cases. The benefits of adapting heritage-listed buildings are the cultural and social values embodied within the building are retained for the wider benefit of the community (Ball 2002; Snyder 2005). Some of the case buildings date from 1897. Though subject to numerous adaptations in order to ensure that market needs are met over time, they are highly regarded by tenants and owners and perceived to embody qualities such as a sense of history and quality and convey a sense of prestige and distinction to occupiers. In the 1200 Buildings Programme, there is a preference for non-listed building adaptations.

4.5.10   Number of Storeys

When the number of storeys is examined, 70% are to buildings under 20 storeys or low- to medium-rise stock. Adaptations to high-rise buildings (i.e. 21–45 storeys high) occurred at 30%. There are some similarities to Wilkinson (2011), where 30% of adaptations occurred to high-rise stock. Higher rates are noted in the low- to medium-rise category, and this is due to the overall size of buildings that have joined the 1200 Buildings Programme.

4.5.11   PCA Grade

The Property Council of Australia building quality grade can be interpreted as an economic factor. Broadly associated with Property Council of Australia building quality grade is a building's capital value, investment value and yield. The economic goal or financial drivers of adaptation cited by a number of 1200 Buildings Programme owners were to increase value post-adaptation, after construction and development costs are taken into account. Adaptation has to be economically viable to be successful (Kincaid 2002), and there has to be market demand to bring about economically viable project. Positive user demand was important in successful building adaptation (Ball 2002) and with vacancy rates for offices being historically low during the period covered by this research; there has been positive user demand throughout the CBD. However, there is another way of interpreting Property Council of Australia building quality grade and that is as a measure of building quality or building amenity levels.

Depending on the condition of a building, it is possible to increase the overall quality with adaptation (Kersting 2006). Office building quality is measured in various ways but, generally and across all land uses, can be stated to be the provision of either a greater number of amenity features and attributes or a higher standard of services, features, fixtures and fittings. In Australia, offices are graded Premium, A, B, C and D grades having progressively less amenity and quality and less capital and rental values (PCA 2007). It is possible to increase the office quality grade from one band to another and increase the rental and capital values. Overall the results showed that ungraded stock was most likely to be worked on (50%), followed by B-grade stock (20%). Premium-, A- and C-grade stock represented 10%, respectively. It appears that owners are most active in working on ungraded followed by B-grade stock. These results contrast to the earlier study (Wilkinson 2011) where B-grade buildings were most likely to be worked on (27%), followed by ungraded stock (24%) and A-grade stock (21%). From 1998 to 2008 owners were most active in adaptation to A- and B-grade stock. Premium stock accounted for 10% of work and reflects the age and condition of this type of stock within the CBD. Half of the projects occurred to offices that are not classed under the Property Council of Australia building quality matrix. This stock is low quality; hence, it does not achieve the standards required for inclusion in the building quality matrix.

4.5.12   Attachment to Other Buildings

Attachment to other buildings refers to the ease of access for contractors to a property. In the CBD many smaller low-rise buildings tend to be attached on two sides, with the larger high-rise stock more likely to be detached. Isaacs (cited in Baird et al. 1996) noted that the attachment to other buildings affected the ease of, and the desirability of, adaptation. Detached buildings are easier to adapt externally because owners can get access to elevations for construction works such as re-cladding the envelope. Internal adaptations are easier with detached or less attached properties as owners can gain access for materials delivery and removal of waste without disturbing or engaging in negotiations with neighbours. Wilkinson (2011) found this to be true in Melbourne CBD adaptations from 1998 to 2008. Buildings that are detached were equally most likely to undergo adaptation (50% of all 1200 Buildings Programme cases). However, buildings attached on three sides accounted for 50% of adaptations that is a marked change to Wilkinson's 1998–2008 study where only 8% of adaptation occurred to such stock. The negative impact of access issues faced by owners in adapting buildings with a high level of attachment to adjoining properties was not found in this sustainable adaptation study.

4.5.13   Site Access

Site access is crucial in building adaptation (Gann and Barlow 1996; Remøy and Van der Voordt 2006). The reason is that contractors need to set up site accommodation and deliver materials and equipment to the building during adaptation. The ease with which this can be accomplished affects the cost and the duration of the adaptation project. Furthermore, site access to and from a building determines whether owners can undertake adaptation with occupants in situ. In this study buildings with least good levels of site access (street only) are most likely to be adapted (50%), followed by buildings with street, side and rear access (30%) and lastly buildings with access on all sides (20%). Collectively half have 'good' to 'very good' access, thereby supporting the assertions of previous studies in respect of accessibility (Isaacs in Baird et al. 1996). Buildings with site access from the street only had a lot of work undertaken, and the issues regarding access for contractors were overcome in some cases by working outside normal working hours at night and weekends which added to time and costs.

4.6   Conclusion

This chapter used ten cases to illustrate how sustainable building adaptation is manifesting itself in the City of Melbourne within the 1200 Buildings Programme. Through the cases it has been possible to gain a deeper understanding of sustainable building adaptation. The case studies showed sustainability measures were 61% related to services, 73% related to energy efficiency, 11% were water economy measures, 17% of measures were to the fabric and 6% had a social sustainability component. When comparing current practice to identify similarities and differences in approach to adaptation in the 1200 Buildings Programme, the findings were that owners are motivated by different drivers; fringe locations feature much more prominently in the cases; aesthetics is important with 60% ranked as attractive; buildings having a wider range of location of services are being adapted; a wider range of land uses are drawn into adaptation than previously; buildings with smaller floor areas are generally adapted; building widths are mostly in the lower range; there is a preference for non-listed buildings; higher rates are in the low- to medium-rise category and ungraded buildings were most likely to be worked on (50%), followed by B-grade stock (20%); detached buildings were equally likely to undergo adaptation (50% of cases) along with buildings attached on three sides (50% of cases); and half of adaptations have 'good' to 'very good' site access.

The implications are that owners and designers should be more realistic of a building's life cycle and avoid over-specification of commercial office buildings whereby extra resources are committed to buildings to provide a hypothetical life of 100 plus years, for example, when in reality societal tastes, needs, perceptions and expectations will have changed so much so that the building will be perceived to no longer meet market expectations. Over-specification in office buildings was found to be a common practice in speculatively designed office buildings in an international comparative study (Wilkinson and English 1998). It appears the practice continues and the problem is exacerbated by regular churn of tenants in commercial stock. If offices are typically over-specified, that is, the design goes beyond what is required within legislation to satisfy the perceptions of the marketplace, additional resources are used and greenhouse gas emissions emitted to deliver the specification. Subsequently, a reduced life cycle follows, whereby components designed to last 50 years or so are replaced within a much shorter time frame, say, 15 years. In an age where sustainability is so often the stated goal, it is imperative that unnecessary waste is avoided wherever possible.

In conclusion, the case studies presented in this part of the text represent an intensive examination of building adaptation undertaken in respect of a defined mature property market, in this case the Melbourne CBD. Importantly, the study coincides with the development of the 1200 Buildings Programme, a City of Melbourne initiative to deliver 38% reductions in building-related greenhouse gas emissions by 2020. This is an initiative that is being replicated to lesser or greater degrees within many cities around the world. The lessons that may be applied from this new knowledge are therefore timely and much needed. The sustainability agenda and the drive to reduce the environmental impact of existing buildings increase the need for effective, soundly reasoned and informed decision-making in respect of adaptations, and this research has made some inroads in this respect. While the environmental attributes were not found to be important in explaining variance in building adaptation in the PCAs between 1998 and 2008, this has changed in the study reported here and more change is likely to occur. Legislation such as Mandatory Disclosure is having an impact in the marketplace with regard to energy efficiency, and this may be replicated in European market with Energy Performance Certificates. As the breadth and depth of sustainability measures is added to the building code, so too will the need to take these attributes into account increase.

In addition, with the roll-out of the 1200 Buildings Programme and the increasing number of exemplar adapted buildings, such as 500 Collins Street, environmental attributes will acquire greater importance and status. Deferred obsolescence and the environmental, technological, social and economic benefits discussed in the research make adaptation an increasingly attractive option to building owners. The capacity to make an informed decision on the timing and the type of adaptation to make and, on which building, is needed to contribute to the challenge facing humankind in respect of mitigating global warming and climate change. Built environment professionals and policymakers need to be conscious of the relationships between all levels of sustainable building adaptations and property attributes in order that they might make more informed choices in their professional life, for the effects of their decisions affect us all.

References

Arge, K. (2005) Adaptable office buildings: theory and practice, *Facilities*, 23(3–4), 119–127.

Arup (2008) *Existing buildings: survival strategies*, Melbourne: Arup.

Australian Government Department of Climate Change and Energy Efficiency. (2010). Available at http://www.climatechange.gov.au/climate-change/greenhouse-gas-measurement-and-reporting/australias-emissions-projections/australias. Accessed on 2nd September 2013.

Australian Institute of Urban Studies. (2007) Environmental indicators for metropolitan Melbourne: bulletin 9 2006/2007. Available at http://www.aius.org.au/indicators. Accessed on 11 August 2013.

Baird, G., Gray, J., Isaacs, N., Kernohan, D. and McIndoe, G. (1996) *Building evaluation techniques*, New York: McGraw-Hill.

Ball, R.M. (2002) Re use potential and vacant industrial premises: revisiting the regeneration issue in Stoke on Trent, *Journal of Property Research*, 19, 93–110.

Brand, S. (1994) *How buildings learn: what happens after they're built*, Harmondsworth: Penguin.

Carbon Planet. (2011) Greenhouse gas emissions by country. Available at http://www.carbonplanet.com. Accessed on 11 August 2013.

Chudley, R. (1981) *The maintenance and adaptation of buildings*, London: Longman Group Limited.

Douglas, J. (2006) *Building adaptation* (2nd edition), London: Elsevier.

Gann, D.M. and Barlow, J. (1996) Flexibility in building use: the technical feasibility of converting redundant offices into flats, *Construction Management and Economics*, 14(1), 55–66.

Highfield, D. (2000) *Refurbishment and upgrading of buildings*, London: E & FN Spon.

Jones Lang LaSalle (2008) Sustainability now an investor priority, press release, Jones Lang LaSalle, Sydney, Australia.

Kersting, J.M. (2006) Integrating past and present: the story of a building through adaptive reuse, Master's Thesis, University of Cincinnati, OH.

Kincaid, D. (2002) *Adapting buildings for changing uses: guidelines for change of use refurbishment*, London: Spon Press.

Lorenz, D., Heard, B., Hoekstra-Fokkink, L., Orchard, J. and Valeri, S. (2008) *Towards a city of Melbourne climate change adaptation strategy: a risk assessment and action plan*, Melbourne: Maunsell Australia Pty Ltd.

Markus, A.M., ed. (1979) *Building conversion and rehabilitation: designing for change in building use*, London: Newnes-Butterworth.

Melbourne City Research. (2006) *City index report 2005: progress against city plan performance indicators*, Melbourne.

Ohemeng, F. (1996) The application of multi-attribute theory to building rehabilitation versus redevelopment options, in proceedings of COBRA, RICS, Las Vegas, NV.

PCA (2007) *Benchmarks survey of operating costs: Melbourne office buildings*, Melbourne: Property Council of Australia.

PCA (2008) *Existing buildings/survival strategies: a toolbox for re-energising tired assets*, Melbourne: Property Council of Australia.

Remøy, H. and Van der Voordt, T.J.M. (2006) A new life: transformation of vacant office buildings into housing, in proceedings of CIBW70 Trondheim International Symposium, Norwegian University of Science and Technology, Trondheim, Norway, June 12–14.

Robson, C. (2003) *Real world research: a resource for social scientists and practitioner-researchers*, Oxford: Blackwells.

Snyder, G.H.. (2005) Sustainability through adaptive reuse: the conversion of industrial buildings, Master's Thesis, University of Cincinnati, OH.

Swallow, P. (1997) Managing unoccupied buildings and sites, *Structural Survey*, 15(2), 74–79.

Szarejko, W. and Trocka-Leszczynska, E. (2007) Aspect of functionality in modernization of office buildings, *Facilities*, 25(3), 163–170.

Wilkinson, S.J. (2011) The relationship between building adaptation and property attributes, PhD Thesis, Deakin University, Burwood, Australia.

Wilkinson, S.J. (2012) The increasing importance of environmental attributes in commercial building retrofits, in proceedings of RICS COBRA, Las Vegas, NV, September 11–13. Available at http://www.rics.org/au/knowledge/research/conference-papers/cobra-2012-environmental-attributes-in-commercial-building-retrofits/. Accessed on 19 August 2013.

Wilkinson, S.J. and English C. (1998) An investigation into the apparent over-specification of speculative UK commercial office buildings: a comparison between the UK, France, Germany, North America, Australia and Japan, in proceedings of RICS COBRA, Las Vegas, NV, September 1998.

Zunde, J. (1989) *Design technology*, Sheffield: PAVIC Publications.

Part II Adaptive Reuse

The author for this part is Dr Hilde Remøy. Hilde is Assistant Professor of Real Estate Management at the Faculty of Architecture, Delft University of Technology, Delft, the Netherlands. She has experience with adaptive reuse from both practice and academia.

The research described in this part is the result of studies undertaken in the period from 2005–2013. Hilde's research focus is adaptive reuse of existing buildings that have lost their original function, related to obsolescence and vacancy of existing buildings and locations. In research and education, she is working on studies concerning the influence of physical property characteristics on obsolescence and adaptive reuse potential and studies to define the future value of reused buildings and cultural heritage. Hilde is the author of several books/book chapters.

This part of the book deals with adaptive reuse. Adaptive reuse as a means of coping with building obsolescence is explored through a review and synthesis of the relevant literature. Based on empirical data, adaptive reuse is discussed and compared to other possible strategies for coping with obsolete buildings; demolition and rebuild, adaptation or consolidation. This part further focuses on structural vacancy of office buildings as an important contributor to office building obsolescence and one of the drivers for adaptive reuse. Typically, most former studies were concerned with one specific topic of adaptive reuse, whereas this research has taught us that taking a broad perspective looking into physical characteristics, involved actors and the assessment method and organisation are important to understand adaptive reuse. Henceforth, the critical success and failure factors for adaptive reuse are identified and explored through case studies. The success and failure factors include market,

Sustainable Building Adaptation: Innovations in Decision-Making, First Edition.
Sara J. Wilkinson, Hilde Remøy and Craig Langston.
© 2014 John Wiley & Sons, Ltd. Published 2014 by John Wiley & Sons, Ltd.

location and building criteria, and appear important in assessing the adaptive reuse potential of existing buildings, and hence contribute to decision-making for adaptive reuse. In assessing the adaptive reuse potential of existing buildings, the market value and use value are critical to financial feasibility. When assessing the adaptive reuse potential of cultural heritage, however, the cultural historic value also has great impact, and may be the key to successfully redeveloping cultural heritage for new use.

Chapter 5 commences with a definition of adaptive reuse and conversion, and gives a broad review on existing adaptive reuse theory. Adaptive reuse has been studied and described internationally, from different perspectives and disciplines. This chapter combines these different studies, developing a theoretical framework for adaptive reuse. One of the main drivers for adaptive reuse, next to common adaptation drivers, is building or functional obsolescence. Functional obsolescence is caused by societal, economic and technological changes, based on which the building is no longer suited to accommodate its original function. Overproduction and under occupancy of offices is a specific cause for obsolescence that is discussed in this chapter.

Chapter 6 describes adaptive reuse as one of four possible strategies for dealing with functional obsolescence and structurally vacant office buildings, compared to three other possible strategies: consolidation, (within use) adaptation and demolish and new build. Different methods and tools for assessing the adaptive reuse potential of obsolete buildings are described and evaluated. The criteria included in assessment tools mainly have an influence on the financial, social or environmental feasibility of the conversion. Therefore, the specific criteria included in the tools are explained. Finally, based on a case study of structurally vacant offices in Amsterdam, the influence of the different criteria on the financial feasibility of the four different scenarios is discussed.

Chapter 7 describes fifteen Dutch case studies of *ex post* office buildings that were converted into housing. First, each case study is described, before the success and failure factors of the different cases are analyzed and compared to each other in a cross case analysis. Based on the comparison of success and failure factors, the risks and opportunities of adaptive reuse projects are explained. Second, this chapter develops a check list for opportunities and risks of adaptive reuse projects. Third, the lessons learnt from the cross case analysis show how opportunities and risks of conversion are closely related to the physical property characteristics. Finally, the adaptive reuse potential of common office types is described.

Chapter 8 focuses on the adaptation of cultural heritage. In adaptive reuse decision-making, all criteria can comprise financial, social and environmental aspects. However, when dealing with cultural heritage, other criteria come into play, or are assessed with a different weighting. Heritage values may be seen as societal values, but using cost-benefit analysis these values can also be monetarised. With adaptive reuse of heritage buildings, the value of cultural heritage is defined more specifically. The experience value of cultural heritage is very important in adaptive reuse. This chapter describes how the experience value of cultural heritage can ascribe a new brand image to the whole area surrounding it, illustrated by case studies.

5

Building Obsolescence and Reuse

Adaptive reuse is defined as a major change of a building with alterations of both the building itself and the function it accommodates. Such conversion is not a new phenomenon; if historical data are consulted, they reveal that conversion has taken place at any place and at all times, internationally and on different scales, contributing to today's much loved historical cities and buildings. Two famous examples are the amphitheatre in Lucca, Italy, from the second century, and the canal houses in Amsterdam from the seventeenth century (see Figure 5.1). The amphitheatre has gone through several transformations and adaptations, and now 1900 years later, what is left of the original structure is merely its spatial configuration; the theatre scene is now a piazza, and the buildings around the piazza have taken the place of rows of seats. The Amsterdam canal houses are quite young compared to the theatre in Lucca, and here next to the spatial configuration, also the images of the facades and heights of the buildings are kept, though these were also adapted several times. The functions of the buildings have also changed a number of times together with the interior floors and the rear facade of the buildings. In this chapter, the theoretical framework for adaptive reuse is drawn and the relationship between adaptive reuse, obsolescence and building lifespan is explained.

The drivers for building conversion are social, economic and environmental. One social driver is the renewed appeal for city centre living and planning policies that reinforce this interest (Heath 2001; Beauregard 2005). There are issues where cities face periods where obsolete buildings blight areas socially (Bryson 1997). Adaptation promotes urban intensification, retains embodied energy and encourages the use of public transport.

Sustainable Building Adaptation: Innovations in Decision-Making, First Edition.
Sara J. Wilkinson, Hilde Remøy and Craig Langston.
© 2014 John Wiley & Sons, Ltd. Published 2014 by John Wiley & Sons, Ltd.

Figure 5.1   The former amphitheatre in Lucca and the canal houses in Amsterdam.

Economic drivers comprise the economic benefits of heritage (Ruijgrok 2006) and the reuse of obsolete buildings. Internationally, reuse of obsolete and redundant office buildings is an important driver (Remøy 2010). A Hong Kong study of the impact of refurbishment on high-density residential property showed a 9.8% increase in value compared to identical un-refurbished property (Chau et al. 2003). In Amsterdam, adaptation is driven by surplus office stock where older buildings are vacated for new buildings, and vacancy concentrates in the older stock where obsolescence occurs. In Australia, environmental drivers are possibly more important and building adaptation is seen as a vital part of sustainable development, allowing a glimpse of the past and imparting character and identity to precincts while referring to their local history (DEH 2004).

Obsolescence is perceived as a problem of economic and social decay. Uncertainty and social insecurity are manifest as vandalism and graffiti, break-ins and illegal occupancy. Investors can spread the risk of obsolescence by building a diverse portfolio and only face the issue of depreciation when selling; however, owners of long-term vacant office buildings suffer a lack of income. High vacancy rates hit investors indirectly due to the negative influence on the market and the negative externalities of vacant offices nearby (Koppels et al. 2011), although investors tend to see the problem as somebody else's problem (Remøy and Van der Voordt 2007a). The investment market is stratified, with new offices procured mostly by institutional investors who sell off older properties where vacancy concentrates to smaller or private investors; such movement is an example of Atkinson's sinking stack theory in action (Langston and Shen 2007).

Office vacancy rates worldwide have been increasing since the 2008 global financial crisis and this led the Dutch Government to acknowledge that long-term vacancy is an issue in the real estate market. While new office buildings increase the urban footprint, older properties remain vacant, occupying scarce land (Chandler 1991; Ball 2002; Remøy 2010). Adaptation of existing offices is a sustainable way of addressing vacancy, either through conversion or within-use adaptation. There is a twofold benefit with office conversions that lower vacancy rates and enhance the sustainability of the built environment by reducing embodied energy in converted residential stock.

Conversion is a way of coping with structural vacancy of office buildings. In the Netherlands, the scale on which since 2001 office buildings have lost their function is so far unprecedented; at the end of 2012 about seven million square metres, equalling 15% of the office space in the Netherlands, was vacant. Due to the large surplus of office buildings in the market, maintenance and adaptation of buildings for their existing use or adaptation into a similar use is not economically or socially viable. Residential conversion of office buildings is the main issue of this chapter. There are several reasons for focusing on this specific functional conversion: first, specific typological aspects of office buildings, like bay width, match the requirements of housing; second, the locational qualities of both functions are often aligned; and third, in most contexts the two functions represent the largest component of real estate markets in volume.

5.2   Conversion Research Worldwide

Research on conversion of redundant or obsolete (office) buildings into housing has been conducted in several academic institutions worldwide and was described in several publications (Gann and Barlow 1996; Tiesdell et al. 1996; Coupland and Marsh 1998; Heath 2001; Beauregard 2005; Langston et al. 2008; Wilkinson et al. 2009; Bullen and Love 2010; Remøy and van der Voordt 2007), describing conversions in London, New York, Toronto, Tokyo, Hong Kong, Australia and the Netherlands. The studies all show conversions of redundant office buildings in central urban areas or Downtown locations.

The cases of London and Toronto are described by Heath (2001), describing popular office to residential conversions as a very successful means of redeveloping inner cities during the 1990s. Both the City of London and the Toronto City core were areas characterised as office districts with little housing that experienced an exodus to the suburbs at 6.00 pm. As the offices were ageing and becoming obsolete, the opportunities for conversion arose. Office building booms in the late 1980s and an economic recession in the early 1990s resulted in a large stock of vacant offices and a dramatic reduction in rents, resulting in a replacement market where tenants moved to newer accommodations at comparable rents (Barlow and Gann 1995). The planning authorities in Toronto and London reacted quite differently to the vacancy problem. While in Toronto the planning system played a key role in bringing forward developments, the London planning system was supportive though not proactive. In Toronto, in the early 1990s, 9000 dwellings were added to the Downtown area. By 2000, the impetus for conversions had slowed down; the office vacancy had fallen to 9%, the most suitable buildings were already transformed, and since the Downtown now had a strong residential market, many obsolete buildings were demolished and new residential accommodation was constructed. While conversions in Toronto were concentrated to the Downtown area, in London conversion was taking place more dispersed, in the different boroughs and in the City of London. The driver for conversions in London was the opportunities that occurred as office rental values fell below those of residential accommodation. The triggers and obstacles of conversion in Toronto and London were numerous; the triggers included demographic and household compositions, changing attitudes and housing demand, causing city centre living to become more popular. In addition, there was little or no demand for the vacant office space from existing or alternative uses. However, the most important factor was the rent gap between the functions offices and housing; as by 1994 in some situations the return on housing was estimated to be 90% higher than that for commercial rented property. In London, the effect increased by office owners refusing to accept the lower rents and thereby contributing to increased obsolescence. The five major triggers and obstacles to the conversion process were found to be physical/design aspects, location, financial/economic aspects, demand and legal aspects.

In the 1960s, the erosion of Lower Manhattan as a business centre signalled the starting point for the government to invest in the quality of the area. A world trade centre, improved public transport, the 'Alliance for Downtown' by corporations located there and the reintroduction of middle-class housing were the four issues that were incorporated in the government's plan. The aim was to improve the Downtown office market. It partly succeeded, though in the late 1980s offices relocated out of Manhattan as a reaction to the economic slump, part of a cyclic development comparable to that of London and Toronto; after an economic boost in the late 1980s, the recession in the early 1990s left a huge amount of Downtown office buildings structurally vacant (Barlow and Gann 1993). Between 1992 and 1995 the office vacancy rates were about 20%. Office tenants who still preferred Manhattan moved to Midtown, as the buildings there were newer, bigger and of a better quality. A large amount of office buildings Downtown was obsolete. As a reaction, in 1995 the New York City Government initiated the Lower Manhattan Revitalisation Plan to enable and subsidise the conversion of obsolete Lower Manhattan office buildings into apartments (Beauregard 2005). Subsidies were given for the conversion of office buildings completed before 1975. The government focused on conversions into studios and small apartments for first-time renters, though the converted offices were also popular with other groups, as the rents were kept relatively low because of the subsidies. However, the area lacked basic services seen as substantial for families or the elderly. The triggers of the successful conversions in Downtown Manhattan were the tight housing market and a high supply of obsolete office buildings. From 1995 to 2005 more than 60 office buildings were converted, and the number of inhabitants in the area grew. Still, the worker population in Lower Manhattan is three to four times larger than the resident population, and there are few services and facilities for residents.

In Australia on the other hand, the take on building adaptation and conversion often concerns sustainability. Upgrading the existing building stock to improve sustainability and reduce CO_2 emissions is a target for the City of Melbourne (Wilkinson and Remøy 2011), along with increasing the proportion of residential population in the CBD. This vision is shared by Perth in Western Australia (Bullen 2007) and other cities in the region, where high vacancy of office space and increased residential construction add to the interest in building conversion.

The same issues are at stake in Hong Kong. Its dense structure offers little space for new developments, and so changes in the urban fabric occur as adaptive reuse or demolition and new construction. With new construction contributing only 2% per year to the building stock, it would take Hong Kong up to 100 years for energy-efficient strategies of new building construction to contribute to reduce energy use and greenhouse gas emissions according to the targets of the Hong Kong Government. Hence, adaptive reuse is needed to reach the goals (Langston et al. 2008).

In Tokyo, some developments were found that are equal to the New York developments. As the office market was climbing up from the recession in the 1990s, new office buildings were added to the market. However, the take-up of offices lagged behind and had not yet recovered as the dot-com crisis hit the market in 2002–2003. Older and smaller office buildings located in secondary streets were becoming obsolete, and conversion has been taking place, though at a smaller scale (Ogawa et al. 2007). Different from New York, albeit the value of the existing buildings is low, the tenancy perspectives for new and large office buildings are still good. Therefore, demolition with new construction has in general been a more interesting option than conversion, often resulting in an increase in scale of the urban fabric. The local government has had little influence on the urban developments, though recent focus on conservation of the urban fabric and urban sustainability might enhance the opportunities for building adaptation and conversion (Minami 2007).

The availability of obsolete office buildings and a tight housing market are found to be the most important triggers for conversion (Heath 2001; Beauregard 2005). This also goes for the Netherlands. Conversion of structurally vacant offices into housing may contribute to increasing and broadening the housing supply and at the same time create possible new use for functionally obsolete office buildings. The conversion potential of structurally vacant office buildings into housing in this market depends on the financial feasibility of conversion.

The appraised market value of office buildings is normally based on the income approach, described by the potential rental income. However, structurally vacant office buildings generate no income and have no perspective of future tenancy. Still, appraisal of structurally vacant office buildings is in most literature (Hendershott 1996; ten Have 2002; Hordijk and van de Ridder 2005) based on potential tenancy of the property, even in situations of market disequilibrium such as studied by Hendershott (1996), using either the cap rate or discounted cash flow methods calculating the net present value (NPV).

The accounted value of structurally vacant office buildings is found too high for redevelopers, who calculate land and existing building value residually, as what is left over after subtracting the conversion costs from the total estimated yield of the redevelopment. This value is found by owners of structurally vacant office buildings to be too low. As long as these two ways of calculating the value of structurally vacant office buildings are not compatible, the price of structurally vacant office buildings will be experienced as too high by redevelopers. In specific markets such as Manhattan, New York, the price per square metre housing is higher than the price per square metre office, and the office market comprises several functionally obsolete office buildings that would need radical adaptation in order to be let out again as offices (Beauregard 2005).

5.3   Building Lifespan and Obsolescence

According to the life cycle perspective on buildings, the building is seen as a cyclical process (Figure 5.2). During the initial phases (initiative, briefing, design and construction processes), the building is created. During the cyclical lifespan, use and operation alternate with adaptations. At certain stages the building will reach a situation where its future usability and value will have to be assessed, and obsolescence may be indicated. This can happen because of the building's technical or functional characteristics (technical or functional obsolescence) or because the costs of use exceeds the benefits of occupation (economic obsolescence). At this point, the building can face major adaptation or its lifespan may be ended: the building may be demolished (Blakstad 2001; Vijverberg 2001; Heijer 2003). The building's lifespan is hence closely connected to the state of obsolescence of the building. The different types of lifespan are explained in Section 5.4.

5.3.1   Technical Lifespan

The technical lifespan is the length of time during which the real estate object can meet the necessary technical and physical demands that are needed to be able to use the building and protect the safety and health of the users. Technical lifespan can be prolonged by building maintenance. Maintenance is defined as repairs that are needed to ensure or restore the original functionality of the building, but does not include measures that improve the initial technical quality of the building.

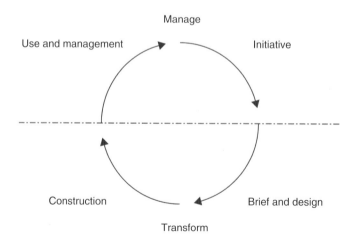

Figure 5.2   The cyclical process of building.

5.3.2   Functional Lifespan

The functional lifespan is the period of time during which a real estate object complies with the functional demands of the user. Next to functionality of use, functional demand relates to aesthetic, social, legal and environmental aspects. This lifespan is ended when the building limits the use. This is linked to the type of use that is located in the building and is therefore also dependent on the specific user of the building. When the end of the functional lifespan is reached, there are several options. An investment can be made to adapt the building, after which it is again able to support the same use as before. Also, the choice can be made to convert the building to accommodate a different function. The environmental lifespan is partly defined as an aspect of the functional lifespan, partly as a parallel to the economic lifespan.

5.3.3   Economic Lifespan

The economic lifespan is the period of time during which the real estate object generates more income than costs. This is the period in which the present value of all future income is higher than the present value of all future costs. The income a property can generate depends on the price, quality and competition in the market; the costs depend on what is needed to maintain the building. The economic lifespan ends when an owner can no longer see a possibility to generate more income than costs. Therefore, the economy can have a very profound effect on the economic lifespan and shorten or lengthen it considerably. The environmental lifespan somehow parallels the economic lifespan, as it describes the time span after which demolition and reconstruction becomes environmentally more favourable than adaptation and reuse. The calculation is based on the environmental load instead of actual costs as a decisive criterion (Van den Dobbelsteen 2004).

These three types of lifespan are interrelated. For example, if the functional lifespan has ended, this usually implies that the economic lifespan also ends. If the functional lifespan ends, it is not possible to find a tenant for the building that means the building can no longer generate income to cover the costs. The end of the functional lifespan may be caused by the end of the technical lifespan; however, it is often the case that a building is still in a technically good condition when the end of the functional lifespan has been reached.

Mismatches between the technical and the functional lifespan of a building frequently occur. The main structure of a building is designed according to current building rules on safety, health, usability, energy use and environment. The main structure of a building can be expected to serve its use for at least 80 years. The functional lifespan of a building can be exterminated through technological progress, causing changes in the user's requirements, influencing both the layout and the facilities offered in new buildings (Baum 1993; Blakstad 2001). In fact, a building can be functionally outdated already after its first lease period. When the building's functional lifespan is ended, the building is functionally obsolete. If the building's technical lifespan is not concurrently ended, there is a mismatch between the building's technical and functional lifespan. This structural vacancy finally also ends the economic lifespan of a building. The reason that a building remains vacant depends on how it compares to the other available buildings on the market. When a building is closer to the end of one of these lifespans than a competing building, it is less likely to attract a new tenant.

5.4   Obsolescence and Vacancy

As stated by Brand (1994), buildings are processes, not completed products. Still, buildings are static and immobile. On the other hand, accommodation strategies are directed towards a specified future match. The future fitness of use of a building is assessed at certain points in time, both in corporate real estate management and in real estate management, and may lead to decisions of selling, demolishing, adapting or converting a building in order to achieve a new match between demand and supply.

As a result of functional obsolescence, the building does not yield any financial benefit to its owner and is therefore also considered financially obsolete. In literature (Nutt et al. 1976b; Salway 1987; Baum 1993; Blakstad 2001), several forms of obsolescence are discussed:

- Aesthetic (visual) obsolescence, resulting from outdated appearance
- Functional obsolescence, resulting from changing ways of working
- Legal obsolescence, resulting from new legal standards
- Social obsolescence, resulting from image issues and increasing demands by occupiers
- Tenure obsolescence, resulting from disagreements between landlord and occupier
- Structural (physical) obsolescence, by Baum referred to as deterioration
- Financial obsolescence, resulting from misbalance between costs and benefits
- Environmental obsolescence, resulting from environmental changes
- Locational obsolescence, resulting from functional obsolescence and image issues of the location
- Site obsolescence, resulting from misbalance between site value and building value

In this chapter, mainly functional obsolescence, technical obsolescence and financial obsolescence are discussed, as parallels to functional, technical and financial lifespans. Functional obsolescence is seen as real and relative and comprises also aesthetic, legal, social and environmental obsolescence. Though the relationship between aesthetic and functional obsolescence can be discussed, they both occur as results of user dissatisfaction, just like social obsolescence. Legal obsolescence, on the other hand, results from new legal standards. However, legal obsolescence is related to the functionality of the building; hence, it is also comprised as functional obsolescence. Environmental obsolescence is on the one side defined by changing building rules. On the other hand, changing trends and office user preferences for sustainable buildings relate environmental obsolescence to functional obsolescence. Structural obsolescence is in this study referred to as technical obsolescence. Tenure obsolescence is not discussed further here, as mainly the relationship between physical characteristics and user preferences is studied. Site obsolescence should, according to definition, lead to demolition and new construction on the site. Locational obsolescence is implicitly incorporated in this study and could be used to describe buildings that are structurally vacant because of the characteristics of their locations.

While in property investment high quality is understood to improve investment return and reduce risk (Salway 1987; Baum 1993; Baum and McElhinney 1997; Bottom et al. 1998), a lack of quality results in obsolescence. Noting that though all buildings deteriorate and become obsolete as they age not all buildings deteriorate with the same speed, Baum studied the depreciation rate of office buildings to prove that the relationship between quality and depreciation is stronger than the relationship between age and depreciation. Baum (1993) refers to obsolescence as a result of changing and qualitatively increasing user demands and deteriorating supply. The quality of office buildings can be determined in terms of occupier utility and hence utility for investors, whereas obsolescence is a source of risk for investors. Applied to office markets with high vacancy levels, significant differences in quality are expected between occupied office buildings and structurally vacant office buildings.

In studies by Duffy and Powell (1997) and Bottom et al. (1998), inflexibility was found to be important indicators of depreciation in commercial buildings, whereas other studies (Healey and Baker 1987) also included the quality of internal finishes, entrance hall and the external appearance of the building. Baum (1991, 1993) and Baum and McElhinney (1997) relate depreciation to obsolescence and use both terms to imply low quality. Depreciation may result from tenure-specific or property-specific factors, and Baum speaks of site value and building depreciation, where building depreciation is a result of physical deterioration and building obsolescence (Table 5.1).

Table 5.1 Important obsolescence factors in former studies.

Baum	Bottom	Healey and Baker
Internal specification	Control of HVAC	Inflexibility
Configuration	Flexibility of floor plan, IT and power	Internal finishes
External appearance	Sanitary facilities	Entrance hall
Physical deterioration	Interior finishes	External appearance

The investment return and risk are considered when investing in real estate. Several studies have been presented using a hedonic pricing analysis to determine the relationship between the physical characteristics of office buildings and locations and the rent prices of offices as an indicator for the value (Dunse and Jones 1998; Fuerst 2007; Eichholtz et al. 2010), while few studies have focused on the risk of vacancy as a threat to the value of investments. However, as the vacancy rates in European office markets have been rising since the 2008 global financial crisis, the risk of structural vacancy in office buildings is becoming a more important factor in the equation than it used to be. In this situation, the rent prices provide insufficient evidence to determine the value of office buildings.

5.5   Quality and Obsolescence: User-Based Property Assessment

From studying physical characteristics that contribute to the quality or the obsolescence of a property, the characteristics are known of office buildings that are vacant in the current market. Using a Delphi survey (Remøy et al. 2007), office accommodation advisors stated that office organisations prefer office buildings and locations with certain physical characteristics (Table 5.2) that enable the organisation to reach their goals. The results from the Delphi survey showed that vacant office buildings can be described by characteristics that are not preferred by office users. Bottom et al. (1998) presented an approach combining the former discussed expert-based appraisal technique with a user-based appraisal technique, best described as a post-occupancy evaluation (POE). The advantages of using POE are recognised in facility and property management (Preiser 1995; Preiser and Vischer 2005) because of

Table 5.2 Physical property characteristics that influence office user preferences.

Building characteristics	Location characteristics
Car parking	Accessibility by car
Exterior appearance	Status
Layout flexibility	Accessibility by public transport
Space efficiency	Facilities
Comfort	Safety
Interior appearance	Business cluster
Recognisable user	—
Technical state	—
Building facilities	—
Year of construction	—
Security	—
Energy performance	—
Routing	—
Bike parking	—
Commodities logistic	—

the possibility of providing feedback information for proactive management. Combining expert-based and user-based property assessments could further help to understand mutations in the office market, revealing which adaptations should be made to enhance the lettability of office buildings and hence prolong the functional and economic lifespan.

The office user preference for specific locations or buildings changes in the course of time, while the location and building characteristics are more or less static. Typically, one important objective for choosing an office location has for the last 20 years been parking possibilities and accessibility by car, while 50 years ago, proximity to customers and employees was far more important (Louw 1996). Building preferences also change due to the revolutions in office work and organisational forms during the last 50 years; work is more informal and individual, organisations less hierarchical and office hours more flexible. As a result, recent office buildings include informal work and meeting spaces, while 50 years ago the office building was monotonous in its spatial layout (Van Meel 2000). However, buildings are by nature static and not readily adaptable (Brand 1994). This means as they age, buildings are less fit to accommodate the function they were designed for and so become increasingly obsolete.

The criteria that office users apply when searching for a certain building (pull factors) are different from the criteria for leaving the building (push factors). This chapter focuses on the criteria that office organisations apply when searching for new office space. Focusing on the pull factors was a deliberate choice as these can be related to the actual market transactions taking place and the contractual rent for the office space. Vacant office buildings on the other hand remain vacant because no other organisations choose to relocate to those buildings. Organisations act dynamically, while buildings are more static; changes are slowly taking place. It is natural for organisations to move from time to time, for instance, when an organisation grows and needs more space than available in the current building. Hence, it is more interesting to find out why organisations do not relocate to a certain kind of building than to find out which buildings they leave behind.

The study by Remøy et al. (2007) showed that accessibility by car and status of the location were the two most important location characteristics for the preference of (all) office users searching for new rental offices, while parking and the appearance of the exterior are recognised as the two most important building characteristics.

Likewise, the characteristics that relate to vacancy, location level, low status of the location and low level of facilities in the surroundings were found to be the most important reasons, while on building level, an unattractive appearance of the exterior was the most important characteristic. Furthermore, less parking places than the surrounding properties was found to be the second most important characteristic. Organisations searching for new office space first scan the available supply within a defined geographical market. Repellent buildings on unattractive locations are the first to be discarded. A second filter applied to the available supply will filter out badly accessible locations and buildings with fewer parking spaces than the surrounding properties.

5.6   The Physical Characteristics of Structurally Vacant Office Buildings

Remøy (2010) studied the association between structural vacancy and location and building characteristics using correlation analyses, graphical charts and regression analyses. The study was conducted analysing the location and building characteristics of 200 office buildings in Amsterdam. Regression analyses not only makes it possible to draw a picture of the correlation between specific physical building and location characteristics and structural vacancy but are used to predict the value of one variable from another variable or a set of variables. Logistic regressions are used to model the relationship between a dichotomous outcome variable and a set of independent variables, also called covariates or predictors. Working with a dichotomous outcome variable makes logistic regressions more stable with a smaller number of cases than is needed for a stable linear regression model. In this research, the independent variables in the logistic regression model were used to predict the odds of structural vacancy. In the sample of 200 office buildings, 106 had structural vacancy.

The results of this study showed strong associations between physical property characteristics and structural vacancy, both on building and location level, and location and building characteristics were revealed that increase the odds of a building being structurally vacant. The location characteristics that most increase the odds of structural vacancy are mono-functionality, lack of status and lack of facilities in the surroundings. The building characteristics that are most closely associated with structural vacancy are bad external appearance of the building, bad internal appearance and low layout flexibility. Although the study was conducted on the Amsterdam office market, it confirms earlier studies (Baum 1991; Louw 1996; Baum and McElhinney 1997; Duffy and Powell 1997; Bottom et al. 1998; Korteweg 2002) and hence shows its generic implication.

Based on the building characteristics found to increase the odds of structurally vacant office buildings, a typology of structural vacancy can be defined. Typologies and type are much discussed in architecture. About the relationship between type and form, Rossi stated that 'the concept of type is permanent and complex, a logical principle that is prior to form and that constitutes (it)' (Rossi and Eisenman 1985). Kohn and Katz (2002) related office building types to the forces of finance, plan, programme and design, while de Gunst and de Jong (1989) focused on the emergence of the office building as a type. Interestingly, in his work on building types, Pevsner (1976) did not recognise the office building as a type, but spoke of it as a subtype, such as government buildings, banks and warehouses.

The typical commercial office building has a simple shape. In Europe, two types dominate (Kamerling et al. 1997; Reuser et al. 2005). The first type is the tall buildings constructed as floors and columns and stabilised by means of a central core of stairs and elevators shafts, in the older cases with extra stabilisation elements in the façade. The second archetype is the low-rise, rectangular building, also built up of floors with columns and stabilising walls in one or two directions, depending on the floor type. However, especially in the Netherlands in the 1980s, load-bearing façades were used frequently in the 'standard' office building, as office development in this period focused on initial building costs, which in the Netherlands at least are lower for this construction type. A core enclosing the stairways and the elevator shaft is normally located in the centre of the building, while the escape routes are normally found at the buildings ends.

5.6.1   Structure and Floors

In the 1980s, two types of structures and floors became popular: a flat-slab floor on beams along the façade and in the centre of the building, supported by columns, and a hollow-core prefabricated floor on façade columns or portals or supported by the load-bearing façade. Both structures are typically linear (Spierings et al. 2004). The flat-slab floor is monolithic and consists of two layers, one layer being prefabricated and installed as shuttering for and connected to the second layer that is reinforced in situ concrete. This floor type spans up to 10 m and was mainly used in central-core tall buildings. The hollow-core prefabricated floor spans up to 15 m and became popular because of the possibility to span from façade to façade, granting flexibility in the layout of floors, and because of the low initial building costs associated with the floor type. For the two structure and floor types, standardisation was pursued. In the Netherlands, the structural grid was more and more a manifold of 1.8 m, like 5.4 or 7.2 m.

5.6.2   Floor Layout, Building Length and Depth

Until the mid-1960s, offices in the Netherlands were built for the owner–users who commissioned the design of their office buildings. This tradition also affected the design of rental offices. The Northern European employee is assertive. Company profiles are democratic and not hierarchic. In combination with building decrees with a strong focus on daylight access, these factors determine the depth of office buildings. Northern European office buildings are narrow compared to buildings in other countries. In the UK, the American-style deep office plates with air conditioning and raised floors were more common (Van Meel 2000). In European and Scandinavian countries, the required daylight access is expressed in the building decree as the equivalent daylight surface: square metres of daylight that enters through windows or other glass building parts. The required daylight access in office buildings equals a vertical glass surface of minimum 5% of the square metres of usable floor space. For housing, daylight access equalling 10% of the usable floor space is required.

Following the standardisation pursuit in the 1980s, the depth of office buildings was also standardised, and in Europe most low-rise office buildings constructed in this period had a depth of 14.4 m, which was possible to span with a hollow-core floor. Developed for a layout of offices flanking a central corridor, the stairs and elevators were situated on one side of the corridor as well, somewhere central in the building or on one end, with additional emergency exits on the ends of the building.

5.6.3   Façade

The energy crisis in 1973 left its mark on European architecture. Insulation of the facade was until then not normal, but in the 1970s it became a standard part of the façade in Northern European countries. Next to completely glazed curtain walls or strip-window facades, load-bearing façades and floor-to-floor prefabricated elements were used. Insulating and sun-reflecting glazing was developed, and the climate facade and the climate window were introduced. Window frames and curtain walls were mostly manufactured from aluminium. The completely glazed curtain wall was clearly expressing the curtain wall principle (Kamerling et al. 1997). Often, the windows in buildings constructed at the end of the 1970s and beginning of the 1980s were sealed. During the 1980s, though, thought was given to the well-being of office employees, as awareness was raised with regard to 'sick building syndrome'. This term points to the fact that office employees became sick from their work surroundings. Some of the problems were caused by façades without operable windows, and the operable windows were reintroduced to the office building, together with individually adjustable heating, ventilation and sunscreens. By standardising the façade measurements, the exterior form of office buildings became more similar.

5.6.4   Stairs and Elevators

The location of stairways in office buildings varies. Since the 1970s the central stability core with an elevator and possibly staircases and facilities has become standard in high-rise and centrally oriented buildings. Depending on the structure, the core is stabilising in one or two directions. Low-rise office buildings have, depending on their length, one or more entrances and elevator cores and escape routes on the end of the buildings.

5.6.5   Location Characteristics

During the 1970s and 1980s, mono-functional business parks were developed on locations outside the city centre. Locations in the city centres were scarce, the plots that were available for office development were small, and development in these areas was rather expensive, compared to the larger locations on the city's fringes (Kohnstamm and Regterschot 1994). Additionally, there was a focus on accessibility by car, and the locations near ring roads and highway exits were far better accessible than the congested city centres (Louw 1996; Jolles et al. 2003). As the new locations were built for cars, there was no need for facilities for employees; the locations were not meant to accommodate other activities than office work. Albeit the locations were well accessible for cars, the parking facilities were not in all cases developed accordingly. In some of the locations, a parking ratio of 1 parking place per 100 m² of office space was used, a ratio that would work if the location was well accessible by car and public transport. Moreover, the locations were developed focusing on large back offices and headquarters of larger corporations, companies that occupy large buildings and have their own facilities like restaurants for the employees. Still, smaller office buildings were also developed in mono-functional locations, creating a need for facilities that was not responded to.

5.7   Selected Adaptive Reuse Projects

There are countless examples of adaptive reuse projects around the world. The following projects have been sourced from Washington, DC (USA), London and Sheffield (UK) and KwaZulu-Natal and Cape Town (South Africa) as part of field studies carried out in 2012. The field studies were conducted by Dr Jim Smith, Bond University, Australia (photographs provided courtesy of Dr Smith).

Three examples of adaptive reuse are presented from Washington, DC. They comprise the National Building Museum (Figure 5.3), the Wonder Bread Building (Figure 5.4) and Parker Flats at Gage School (Figure 5.5).

Figure 5.3 National Building Museum, Washington, DC, USA. Photographs provided courtesy of Dr Smith.

Figure 5.4 Wonder Bread Building (White Cross Bakery), Washington, DC, USA. Photographs provided courtesy of Dr Smith.

Figure 5.5 Parker Flats at Gage School, Washington, DC, USA. Photographs provided courtesy of Dr Smith.

The National Building Museum was constructed between 1882 and 1887 for the US Pension Bureau. The interior of the building is dominated by an impressive central space used for social and political functions with eight huge Corinthian columns. The building was used as offices until the 1960s and was eventually restored and renovated in 1997 to become the National Building Museum. The Wonder Bread Building shows construction in progress in September 2012 by McCullough Construction. The renovation project redevelops seven industrial buildings within the existing preserved façade into a 9500- m^2 multifunction building comprising new retail and commercial office space. Parker Flats at Gage School was built in 1904–1905 in the high Colonial Revival architectural style and was listed on the National Register of Historic Places in 2008. The School was converted to 92 loft-style flats and underground car parking in 2009.

Seven examples of adaptive reuse are presented from London and Sheffield (UK). They comprise The Tanks at the Tate Modern Art Gallery (Figure 5.6), Blackfriars Bridge and Station (Figure 5.7), Battersea Power Station (Figure 5.8), University of Greenwich (Figure 5.9) and St Pancras Railway Station (Figure 5.10) from London and Hillsborough Barracks (Figure 5.11) and Kelham Island Quarter (Figure 5.12) from Sheffield. The most visited modern art gallery in the world, the Tate Modern Art Gallery, is housed in the former Bankside Power Station built in two stages in 1947 and 1963. The power station closed in 1981 and the art gallery opened in 2000. The three large underground oil tanks and adjoining spaces were refurbished as the gallery expanded and displayed video art from the collection. The original Blackfriars Bridge was built in 1886, and in 2011/2012 the new roof over the bridge featured 4400 solar photovoltaic panel (6000 m^2), rain harvesting systems and sun pipes for natural lighting. The listed Battersea Power Station was built in the 1930s and decommissioned in 1983, since which time it has been vacant. Over the years several proposals have stalled, but redevelopment work should commence in 2013 to build 2013 homes, office space shops and a park on this site. Sir Christopher Wren designed the Old Royal Naval College in the seventeenth century, now the Greenwich campus of the University of Greenwich. Three Schools (Business, Computing and Humanities) have been based in the Old Naval College buildings since 1999. St Pancras Railway Station was originally opened in 1868. The buildings, which were under threat of demolition in the 1960s, were expanded and renovated in the mid-2000s to include not only the railway platforms but a shopping centre, many individual shops, hotel and a bus station. The extensive complex of buildings at Hillsborough Barracks on a 9-ha site was developed from 1848 and completed in 1854. The army used the barracks until 1930 when a large manufacturing chemist occupied the buildings until the late 1980s when it fell into neglect. The first stage of the redevelopment of the buildings was completed in 1991 with the opening of a major supermarket and car park. This was followed by another major store, individual shops, petrol station, a hotel, bar, bus station and community college. Finally, Kelham Island Quarter was formerly an

Figure 5.6   The Tanks at the Tate Modern Art Gallery, Bankside, London, UK.

Figure 5.7   Blackfriars Bridge and Station, London, UK.

Figure 5.8   Battersea Power Station, Vauxhall, London, UK.

Figure 5.9   University of Greenwich, London, UK.

Figure 5.10   St Pancras Railway Station, London, UK.

Figure 5.11   Hillsborough Barracks, Sheffield, UK.

Figure 5.12   Kelham Island Quarter, Sheffield, UK.

industrial area in this major steelmaking city. It has been regenerated into a number of residential and social uses. These include an industrial museum using some of the old premises, chimney house for events, retention of five significant pubs, a new boutique brewery, flats, bars, restaurants, shops and other commercial uses.

Three examples of adaptive reuse are presented from KwaZulu-Natal and Cape Town (South Africa). They comprise Tsonga Shoes from KwaZulu-Natal (Figure 5.13) and Breakwater Lodge (Figure 5.14) and the African Trading Post Building (Figure 5.15) from Cape Town. Tsonga Shoes, now known internationally, is located in an abandoned school building in a small village, and their innovative use of two steel containers for offices, kitchen and workshop provides adequate accommodation for the ladies they employ. Breakwater Lodge was formerly a prison built in 1859 and a distinctive four-turreted building constructed in 1901. Since 1991 the University of Cape Town has converted it into a residential Graduate School of Business, and the Protea hotel group has redeveloped the whole site into a hotel for the university and external guests with restaurants and a bar. Finally, the African Trading Post Building was originally the offices of the Cape Town Port Captain built in 1904. The building now houses a three-level shop for African arts, crafts and furniture since 2001.

Figure 5.13   Tsonga Shoes, Lidgetton, KwaZulu-Natal, South Africa.

Figure 5.14   Breakwater Lodge, Cape Town, South Africa.

Figure 5.15   The African Trading Post Building, Waterfront, Cape Town, South Africa.

5.8   Conclusion

Structural vacancy is a problem to owners as their buildings do not deliver any direct yield. However, structural vacancy is also a societal problem. For instance, locations with several structurally vacant office buildings develop in a downward spiralling movement; the buildings deteriorate and are devaluated, causing financial loss to the owner and to the municipality by lower taxes and land lease incomes. Enhancing conversion possibilities by facilitating legal processes and developing policies on transformations is the task of the municipality. Real estate developers, housing associations and housing investors are potentially interested in transformation of structurally vacant office buildings, though the purchasing price of these buildings is found to be too high by these actors.

The locations of structurally vacant buildings to a great extent determine the building's adaptive reuse potential. Mono-functional office locations are not found suited for housing, unless the location is transformed by adding more housing and facilities to the location. Office buildings with cultural–historical, architectural, symbolic, and intrinsic values or experience value are often successfully converted. The adaptive reuse potential of newer office buildings depends more on financial/economic, functional, technical and legal aspects, influencing the financial feasibility of conversions. If the housing market is tight, housing prices rise and the adaptive reuse potential of office buildings increases. The focus on financial feasibility and revenues is easily explained as the actors in conversion processes are mostly commercial parties. However, to become more sustainable, actors in real estate development and investment should consider conversion more often and weigh financial profit against sustainability goals. Increasing a building's lifespan, through conversion, is a way of achieving a more sustainable built environment. Adaptation or conversion is feasible if the building is functionally, technically and financially fit for new use and if possible complying with the existing legal framework. The building's future value is determined by its value to the user, meaning the value in use by means of supporting the user's main activity and its value to the owner, determined by the return on investment and therefore partly determined by the buildings (potential) value in use to the user. Next to financial motives, conversion may be triggered by a monumental status of the building or by specific historic, cultural or architectural values.

References

Australian Government, Department of the Environment and Heritage. (2004) *Adaptive reuse: preserving our past, building our future*, Canberra: Commonwealth of Australia, Department of Environment and Heritage. Available at http://www.environment.gov.au/heritage/publications/protecting/pubs/adaptive-reuse.pdf. Accessed on 21 August 2013.

Ball, R.M. (2002) Re use potential and vacant industrial premises: revisiting the regeneration issue in Stoke on Trent, *Journal of Property Research*, 19, 93–110.

Barlow, J. and Gann, D. (1993) *Offices into flats*, York: Joseph Rowntree Foundation.

Barlow, J. and Gann, D. (1995) Flexible planning and flexible buildings: reusing redundant office space, *Journal of Urban Affairs*, 17(3), 263–276.

Baum, A. (1991) *Property investment depreciation and obsolescence*, London: Routledge.

Baum, A. (1993) Quality, depreciation and property performance, *The Journal of Real Estate Research*, 8(4), 541–565.

Baum, A. and McElhinney, A. (1997) The causes and effects of depreciation in office buildings: a ten year update, working paper of the Department of Land Management and Development, University of Reading, UK.

Beauregard, R.A. (2005) The textures of property markets: downtown housing and office conversions in New York City, *Urban Studies*, 42(13), 2431–2445.

Blakstad, S.H. (2001) A strategic approach to adaptability in office buildings, PhD Thesis, Norwegian University of Science & Technology, Trondheim, Norway.

Bottom, C., McGreal, S. and Heaney, G. (1998) The suitability of premises for business use: an evaluation of supply/demand variations, *Property Management*, 16(3), 134–144.

Brand, S. (1994) *How buildings learn: what happens after they're built*, Harmondsworth: Penguin.

Bryson, J.R. (1997) Obsolesence and the process of creative reconstruction, *Urban Studies*, 34(9), 1439–1459.

Bullen, P.A. (2007) Adaptive reuse and sustainability of commercial buildings, *Facilities*, 25(1–2), 20–31.

Bullen, P.A. and Love, P.E.D. (2010) The rhetoric of adaptive reuse or reality of demolition: views from the field, *Cities*, 27(4), 215–224.

Chandler, I. (1991) *Repair and refurbishment of modern buildings*, London: B.T. Batsford Ltd.

Chau, K.W., Leung, A.Y.T., Yui, C.Y. and Wong, S.K. (2003) Estimating the value enhancement effects of refurbishment, *Facilities*, 21(1), 13–19.

Coupland, A. and Marsh, C. (1998) The Cutting Edge 1998: the conversion of redundant office space to residential use, RICS Research, University of Westminster, London.

Duffy, F. and Powell, K. (1997) *The new office*, London: Conran Octopus.

Dunse, N. and Jones, C. (1998) A hedonic price model of office rents, *Journal of Property Valuation and Investment*, 16, 297–312.

Eichholtz, P., Kok, N. and Quigley, J.M. (2010) Doing well by doing good? Green office buildings, *American Economic Review*, 100(5), 2492–2509.

Fuerst, F. (2007) Office rent determinants: a hedonic panel analysis, Social Science Research Network. Available at http://ssrn.com/abstract=1022828. Accessed on 8 August 2013.

Gann, D.M. and Barlow, J. (1996) Flexibility in building use: the technical feasibility of converting redundant offices into flats, *Construction Management and Economics*, 14(1), 55–66.

de Gunst, D.D. and de Jong, T. (1989) *Typologie van gebouwen; planning en ontwerp van kantoorgebouwen*, Delft: Delftse Universitaire Pers.

Healey, P. and Baker, A. (1987) *National office design survey*, London: Healey and Baker.

Heath, T. (2001) Adaptive re-use of offices for residential use: the experiences of London and Toronto, *Cities*, 18(3), 173–184.

Heijer, A.C.d. (2003) *Inleiding vastgoedmanagement*, Delft: Publikatieburo Bouwkunde.

Hendershott, P.H. (1996) Valuing properties when comparable sales do not exist and the market is in disequilibrium, *Journal of Property Research*, 13(1), 57–66.

Hordijk, A. and van de Ridder, W. (2005) Valuation model uniformity and consistency in real estate indices: the case of The Netherlands, *Journal of Property Investment and Finance*, 23(2), 165–181.

Jolles, A., Klusman, E., Teunissen, B. and City of Amsterdam Physical Planning Department (2003) *Planning Amsterdam scenarios for urban development 1928–2003*, Rotterdam: NAI Publishers.

Kamerling, J.W., Bonebakker, M. and Jellema, R. (1997) *Hogere bouwkunde jellema Dl 9 utiliteitsbouw; bouwmethoden*, Leiden: SMD/Waltman.

Kohn, A.E. and Katz, P. (2002) *Building type basics for office buildings*, New York: Wiley.

Kohnstamm, P.P. and Regterschot, L.J. (1994) *De manager als bouwheer de rol van de bestuurder bij de realisatie van nieuwe huisvesting*, Den Haag: Ten Hagen en Stam.

Koppels, P., Remøy, H. and El Messlaki, S. (2011) The negative externalities of structurally vacant offices, in *ERES*, Jansen, I. and Appel-Meulenbroek, R. eds., Eindhoven: Eindhoven University of Technology.

Korteweg, P.J. (2002) *Obsolescence of office buildings: problem or challenge?* Utrecht: KNAG.

Langston, C. and Shen, L.Y. (2007) Application of the adaptive reuse potential model in Hong Kong: a case study of Lui Seng Chun, *The International Journal of Strategic Property Management*, 11(4), 193–207.

Langston, C., Wong, F., Hui, E. and Shen L.Y. (2008) Strategic assessment of building adaptive reuse opportunities in Hong Kong, *Building and Environment*, 43(10), 1709–1718.

Louw, E. (1996) *Office buildings and location; a geographical investigation of the role of accommodation in office location decisions*, Delft: Delftse Universitaire Pers.

Minami, K. (2007) A study of the urban tissue design for reorganizing urban environments, in *BSA 2007*, Kitsutaka, Y. ed., Tokyo: Tokyo Metropolitan University.

Nutt, B., Sears, D., Walker, B. and Holliday, S. (1976a) *Theory and applications obsolescence in housing*, Westmead: Saxon.

Nutt, B., Walker, B., Holliday, S. and Sears, D. (1976b) *Housing obsolescence*, Hants: Saxon House.

Ogawa, H., Kobayashi, K., Sunaga, N., Mitamura, T., Kinoshita, A., Sawada, S. and Matsumoto, S. (2007) A study on the architectural conversion from office to residential facilities through three case studies in Tokyo, in *Building Stock Activation 2007*, Kitsutaka, Y. ed., Tokyo: Tokyo Metropolitan University.

Pevsner, N. (1976) *A history of building types*, London: Thames and Hudson.

Preiser, W.F.E. (1995) Post-occupancy evaluation: how to make buildings work better, *Facilities*, 13(11), 19–28.

Preiser, W.F.E. and Vischer, J. (2005) *Assessing building performance*, Oxford: Butterworth-Heinemann.

Remøy, H. (2010) *Out of office: a study of the cause of office vacancy and transformation as a means to cope and prevent*, Amsterdam: IOS.

Remøy, H. and Van der Voordt, T.J.M. (2007) Conversion of office buildings; a cross-case analysis, in *Building Stock Activation 2007*, Kitsutaka, Y. ed., Tokyo: Tokyo Metropolitan University.

Remøy, H., Koppels, P.W., van Oel, C. and de Jonge, H. (2007) Characteristics of vacant offices: a Delphi-approach, in *ENHR 2007*, Boelhouwer, P., Groetelaars, D., Ouwehand, A. and Vogels, E. eds., Rotterdam: Delft University of Technology.

Reuser, B., Van Dongen, E., Gros, P., Van der Varst, R., De Zoete, M., Pulles, T. and Areias, R.T. (2005) *What's in a box, in Reuser, B.*, Delft: Delft University of Technology.

Rossi, A. and Eisenman, P. (1985) *The architecture of the city*, Cambridge, MA: MIT Press.

Ruijgrok, E. (2006) The three economic values of cultural heritage: a case study in the Netherlands, *Journal of Cultural Heritage*, 7(3), 206–213.

Salway, F. (1987) Building depreciation and property appraisal techniques, *Journal of Property Valuation and Investment*, 5(2), 118–124.

Spierings, T.G.M., Amerongen, R.P.v. and Millekamp, H. (2004) *Jellema hogere bouwkunde, Dl. 3, Bouwtechniek, draagstructuur*, Utrecht: ThiemeMeulenhoff.

Ten Have, G.G.M. (2002) *Taxatieleer onroerende zaken*, Leiden: Stenfert Kroese.

Tiesdell, S., Oc, T. and Heath, T (1996) *Revitalizing historic urban quarters*, Oxford: Architectural Press.

Van den Dobbelsteen, A.A.J.F. (2004) The sustainable office an exploration of the potential for factor 20 environmental improvement of office accommodation, PhD Thesis, Delft University of Technology, Delft, The Netherlands.

Van Meel, J.J. (2000) *The European office; office design and national context*, Rotterdam: 010 Publishers.

Vijverberg, G.A.M. (2001) *Renovatie van kantoorgebouwen literatuurverkenning en enquete-onderzoek opdrachtgevers, ontwikkelaars en architecten*, Delft: Delftse Universitaire Pers Science.

Wilkinson, S.J. and Remøy, H. (2011) Sustainability and within use office building adaptations: comparison of Dutch and Australian practices, in proceedings of Pacific Rim Real Estate Society, Gold Coast, Australia, January 16–19.

Wilkinson, S.J., James, K. and Reed, R. (2009) Using building adaptation to deliver sustainability in Australia, *Structural Survey*, 27(1), 46–61.

6

Reuse versus Demolition

Property owners have several options for coping with structurally vacant office buildings (Table 6.1). There are four categories of alternatives: consolidation, adaptation or upgrading, demolition and new construction, and conversion. Most owners of vacant office buildings choose a form of consolidation: to do nothing but to wait for better times. Consolidation includes actions like searching for new tenants and disposal of (or selling) the property. The choice is based on several presuppositions. This chapter introduces several tools and instruments that may be valuable to the assessment of building quality and adaptive reuse potential.

The market value of an office building is based on rent value; hence, the sale of a vacant building yields less than the sale of an occupied building. The building will not be sold in accordance with its book value, which is often based on a presupposed 100% rent for the entire investment period. Real estate owners and investors in that case regard selling a (partly) vacant building as financial loss for the seller. For housing market investors and real estate developers, high asking prices are a reason for not converting vacant office buildings into housing. The different real estate markets are separated; office market actors have little knowledge about the housing market, and vice versa, and they tend to have little affinity for each other's way of thinking. Among the stakeholders on the real estate market, there is a general lack of knowledge about transformation processes. Despite this, at different moments in time, depending on the local market conditions, vacant office buildings worldwide have been converted into housing. Specific examples lie in New York, London, Toronto and Paris in the 1990s and the Netherlands in the 2000s.

Sustainable Building Adaptation: Innovations in Decision-Making, First Edition.
Sara J. Wilkinson, Hilde Remøy and Craig Langston.
© 2014 John Wiley & Sons, Ltd. Published 2014 by John Wiley & Sons, Ltd.

Table 6.1   Market, location and building scale levels for assessing conversion.

Scale	Conversion aspects
Market	Supply and demand, financial, location, building
Location	Legal, functional, cultural, sustainable
Building	Legal, financial, functional, technical, cultural, architectural, sustainable

Another alternative for coping with structural vacancy is adaptation for other office market segments or adaptation of the property. Though smaller adaptations are performed every 5 years (Vijverberg 2001; Douglas 2006), at some point the building might be functionally obsolete, and a more radical intervention is needed. However, in markets with high levels of vacancy or with location obsolescence, there is a risk that the positive effect of adaptation will be less than the costs of intervention.

Next to conversion, demolition and new construction is an intervention that creates possibilities for developing a new building fit to future users' needs and is especially interesting in a declining office market. However, redevelopment takes time and leads to a delay of income and disrupts both market and location development, and if the building is technically in a good state, redevelopment is a waste of resources.

Mothballing a building and temporarily allowing use for housing to avoid the building being illegally occupied are not permanent solutions for coping with structural vacancy but may precede adaptation, redevelopment and conversion or even be seen as part of consolidating. However, mothballing may bring about damages to the building and will imply that repair and redecorations are necessary before the building can be rented out again.

Finally, structural vacancy can be coped with by conversion. Conversion may be expensive and disrupt the incomes from and the use of the building. Its future market value accommodating the new function must be higher than for offices. However, if working out successfully, conversion sustains a beneficial and durable use of the location and building, implies less income disruption than redevelopment and has higher social and financial benefits.

6.2   Decision-Making Criteria

Conversion is a possible development when a building is structurally vacant and is assessed to be functionally or economic obsolete while its technical lifespan is not ended. Table 6.2 lists various options for property owners, along with their benefits and drawbacks. Decision-making criteria for how to cope with obsolete buildings are mostly based on financial aspects. However, sustainability is an increasingly important issue that is taken into account when deciding whether to keep a building or to demolish it. Several tools were developed to assess the capacity of a building for conversion. The tools are mostly concerned with the application of possible new functions (Remøy and Van der Voordt 2007a, b). Though it is not made explicit, the financial feasibility of conversion is implicitly taken into account in the different tools that are developed to assist decision-making about conversion of office buildings.

Table 6.2 Options for property owners for structurally vacant buildings.

Option	Benefits	Drawbacks
Maintain in current state (consolidate)	Preserves the property Sustains existing use Ensures ongoing service and lifespan	Requires maintenance costs though no incomes are generated
New tenancy – better study of the market	Find a suitable tenant, may ensure ongoing beneficial use of the property	May be time consuming to find a user for a structurally vacant building; requires maintenance, refurbishment or incentives
Mothball	Minimises running costs, such as cleaning, heating and lighting	Costly to keep safe and secure; vulnerable to vandalism and squatting, dust and dirt accumulation and dampness in the building; no rental income
Anti-squat	Minimises running costs, secures the building against squatting and vandalism	Exposed to wear and tear, inhabitation may influence possible tenancy negatively
Dispose	Realises asset/site value, reduces management and operating costs	Loss of potentially useful asset, price may not correspond to book value
Demolition and new building	New building tailored to meet users preferences	Disruptive and expensive, delay of income, location characteristics cannot be influenced
Adapt and renovate	Enhances the physical and economic characteristics of the building, delays deterioration and obsolescence, reduces the likelihood of redundancy, sustains the building's long-term beneficial use	Disruptive and expensive, extended lifespan is unlikely to be as great as a new building, upgraded performance cannot wholly match that of a new building, location characteristics cannot be influenced
Convert	Enhances and alters the physical and economic characteristics of the building, prevents deterioration and obsolescence, sustains the building's long-term beneficial use, sustains social coherence in the area	Disruptive and expensive, market uncertainty, location characteristics may not suit new function, building costs may be out of control, new rental function may not be the core business of the owner

6.3   Tools, Scans and Instruments

Based on research results, different tools and instruments have been developed to analyse buildings' conversion potential and the feasibility of conversions (Hek et al. 2004; Zijlstra 2006; Geraedts and Van der Voordt 2007; Hofmans et al. 2007) and may be of use at different stages of the conversion process. Most of the tools were developed as checklists and are based on thorough studies of building conversions. (Van der Voordt et al.

2007). Three of the tools that are discussed were developed at Delft University of Technology: one was developed by a construction management company and the other addressed the architects' approach to conversion projects.

6.3.1   The Transformation Meter

In order to be able to judge office buildings on their potential for transformation into housing, the 'transformation meter' was developed by Geraedts and Van der Voordt (2003, 2007) as a quick scan. This tool consists of criteria to assess the value of a building and its location for housing, based on the physical aspects of building and location and with some criteria considering organisational aspects and market aspects. Some of the criteria are 'veto criteria', meaning that if they have a negative influence on transformation potential, adaptive reuse is unfeasible. Most veto criteria are location characteristics. Depending on the target group, while the transformation of the building can be financially feasible, its locational characteristics cannot be easily changed. The transformation meter has been developed to assist decision-making at the beginning of a possible transformation trajectory.

The vacancy duration is seen as one of the most important criteria before considering transformation as a means of coping with vacancy, and transformation may be advised for buildings with structural vacancy. Also the municipal policy and the zoning plan for the area where the building is situated are taken into consideration. The demand for housing within a specific area is the next issue that is discussed. However, in the Dutch housing market demand is higher than supply, and, especially in the denser areas where most office buildings are located, housing demand is rising.

Assessing the transformation potential of a building using the transformation meter follows five steps. In Step 1, the so-called veto criteria are assessed. These are demand for housing, urban location (considering the two aspects zoning plan and serious public health risk) and dimension of the building structure (considering 2.60 m free height of the floors), and at the organisational level there must be an enthusiastic developer (the building must fit within the developer's portfolio and the owner of the building must be willing to sell or to redevelop). One could argue that the organisational veto criteria are actually superfluous since transformation and using the transformation meter will not be considered if there is no one interested in transforming the building.

Steps 2 and 3 using the transformation meter are a scan of the transformation potential of the building and its location based on gradual criteria (Table 6.3). In these steps, the characteristics of the building and the location are weighted: the location criteria are multiplied by five, being more important than the building criteria, which are multiplied by three. Twenty-three location criteria and 28 building criteria are used.

Table 6.3 Feasibility scan using gradual assessment criteria.

Location	Gradual criterion	Appraisal (yes/no)
Functional		
Urban location	Remote industrial or office park	
	Building gets little or no sun	
Distance and quality of facilities	Shops for daily necessities >1 km	
	Public meeting space (square, park) >500 m	
	Hotel/restaurant/cafe >500 m	
	Bank/post office >1 km	
	Education, sports, basic medical facilities >1 km	
Public transport	Distance to railway station >1 km	
	Distance to bus/metro/tram >250 m	
Accessibility by car and parking	Obstacles, traffic congestion	
Congestion: one-way traffic, no parking, traffic jams	Distance to parking place >250 m	
	<1 parking place per housing unit	
Cultural		
Status of neighbourhood	Situated near city edge (e.g. near motorway)	
	No housing in immediate vicinity	
	Poor green space in neighbourhood	
	Area has poor reputation/image, vandalism	
	Noise or stench (factories, trains, cars)	
Legal		
Urban location	Noise load on facade >50 dB (limit for offices 60dB)	
Ownership of land	Lease	
Building	**Gradual criterion**	**Appraisal (yes/no)**
Functional		
Year of construction or renovation	Office building recently built or modified (<3 years)	
Vacancy	Building vacant <3 years or partly vacant	
Features of new housing	<20 units of minimal 50 m² can be realised	
	Unsuitable layout for selected target group	
Extendibility	Not horizontally or vertically extendable	
Technical		
Maintenance	Poorly maintained	
Structure dimension	Building depth <10 m	
	Structural grid <3.6 m	
	Distance between floors >3.6 m and <5.5 m	
Support structure	In poor condition, not sufficient for housing	
Facade	Not adaptable, impossible to attach interior walls	
	Windows cannot be reused/windows are not operable	
Installations	Impossible to fit vertical ventilation shafts	
Cultural		
Character	Lack of identity	

(continued)

Table 6.3 *(Cont'd).*

Location	Gradual criterion	Appraisal (yes/no)
Access entrance	Unsafe entrance	
Legal		
Environment	Acoustic insulation of floors	
	Poor thermal insulation	
	Too little daylight, less than 10% of equivalent floor area	
	No elevators in buildings higher than four floors	
	No emergency stairways or no sufficient stairways	

Adapted from Geraedts and Van der Voordt (2007)

Table 6.4   The influence level of conversion costs.

High costs, high variability	High costs, low variability
Structure	Inner walls
Facade	Ceilings
Mechanical installations	Electrotechnical installations
Total contractor costs	Fixed interior
Acquiring costs	Low costs, low variability
Low costs, high variability	**Foundation**
Roof	Elevators
Floors	Domain
Stairs, ramps, railings	—

Step 4 in the transformation meter is a financial feasibility scan. This scan uses cost outlines from reference housing projects. This part of the scan is actually not so much a tool but may be used to raise awareness about the effect of the level of intervention on the conversion building costs and is based on a cross-case analysis of 11 converted buildings (Geraedts and De Vrij 2004). The study revealed that changes in the structure, facade, installations, inner walls, ceilings and fixed interior increase the building costs the most, together with the total contractor costs and the purchasing costs. However, interior walls, ceilings, electrical installations and fixed interior costs were considered costs that are always made, whereas changes in the structure, facade and mechanical installations depend on the state of the original building. Geraedts and De Vrij (2004) described these differences as having a low or high influence on the variation in building costs (see Table 6.4).

Finally, the transformation meter is concluded by Step 5, assessing possible risks of the development and construction phase, followed by opportunities for eliminating the risks. If using the transformation meter for assessing the conversion potential of a building, most of the issues in the risk list are already known.

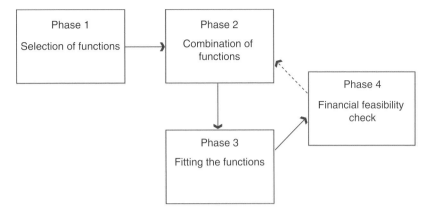

Figure 6.1   The phases of the programmatic quick scan (Hek et al. 2004).

6.3.2   Programmatic Quick Scan

Hek et al. (2004) developed an instrument consisting of four phases (see Figure 6.1). The first phase considers defining possible functions based on the location characteristics and financial, societal, technical and procedural aspects. The study is hierarchical, starting with the location characteristics and then tuning to arrive at a definition of possible functions. In the second phase, combination possibilities of different functions should be studied, starting with possible interaction and synergy effects between the different functions and then developing a concept for fitting the functions in the building. In the third phase the programme is fitted to the building using sketches, and in the fourth phase the financial feasibility of the plan is assessed based on the preliminary sketch plan. The instrument refers to the transformation meter in its use of checklists. In every phase, a checklist should be filled in, assigning scores to decide the feasibility of conversion of the building and the potential for reusing the building for a specific function. Following, the scores are weighted and the feasibility of converting the building into the chosen function(s) is assessed.

6.3.3   Architectural Value

According to Zijlstra (2006), buildings are about the past, the present and the future. Interventions in a building should therefore be preceded by a study of the building's contextual aspects (i.e. original commission, location and architect), next to a study of the building's architecture, in order to decide the potential changeability of the building. More of a method than a tool, this method assumes three levels of time: commencing, ageing and continuing. Within these layers of time, the building elements space, structure, substance and services are studied. The technical lifespan and the technical state of the building are important, as technical decay is often seen as the most important aspect of the ageing of the building, and thus it is also important for the continued life of the building.

Analysing buildings with these aspects in mind, new possibilities are created, offering possibilities for a different way of living, working and recreating. By studying the possibilities before starting the design process, buildings can be kept for continuation instead of being lost to decay. The extra layer of time that existing buildings offer generates urban continuity and additional quality to the functions accommodated in the building (Remøy et al. 2009b).

6.3.4   The Architects' Method

The architects' approach to redesign and redevelopment of a building is not so much a tool or instrument developed for anyone to use, but rather the methodical way of studying and analysing a specific building, a case, in order to conclude the study by making a design for conversion. An understanding of the values (architectural value, value in use, historical and cultural value) of the existing building and an interpretation of these values that makes the building fit for new use are the contribution of the architect to conversion (Coenen 2007). The architects' study in this case needs support by a financial feasibility study and, if applicable, a historical study. The architects' assessment of the building's transformation potential is a kind of retrospective reasoning based on possibilities, departing from ideas of possible future functions and designs for the building, using intrinsic knowledge of typology, construction, space use and dimensions of both the existing building and the possible new functions in order to find a programme that suits the building and enhances the building's architectural quality (Oudijk et al. 2007). The architects' method is, together with a thorough study of financial feasibility, a method often used for assessing the transformation potential of office buildings.

6.3.5   The ABT Method: An Instrument Developed in Practice

ABT is a Dutch multidisciplinary consultancy firm in structural engineering. The firm has contributed to several conversion processes and has developed a quick scan for assessing the conversion capacity of existing buildings, seeing two issues as the most important to assess, that is, possible new functions in the building and the costs of conversion. The ABT quick scan consists of three steps: inspecting the building, controlling (legislation) and valuing (meaning evaluating the technical state, functionality, flexibility, architectonic, historical and 'visual and emotional' quality and assessing five aspects of the building – structure, facade, entrances, fixed interior and installations) and also assessing the condition, legislation and quality of the location (Hofmans et al. 2007). The method is structured as a tree diagram (see Figure 6.2), where the building is central and where the location is seen as a sixth aspect of the building's attributes.

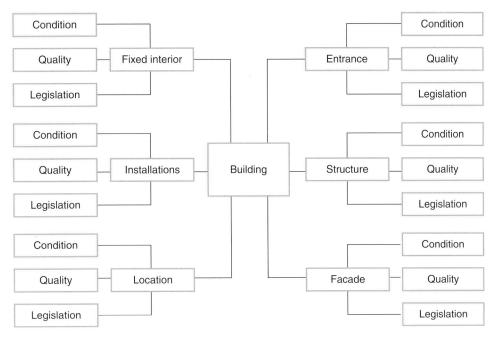

Figure 6.2   Tree diagram (ABT method).

6.4   Decisions-Based on Financial Arguments

Conversion can be compared to other strategies for coping with structurally vacant office buildings, like consolidation, adaptation and demolition with new construction (Vijverberg 1995; Douglas 2006). Monumental, architecturally interesting buildings and buildings with specific visual qualities may be converted despite being maladaptive and despite high conversion costs, because they are either listed monuments, found important by interest groups, seen as interesting acquisition projects by developers or seen as image lenders for a large development. However, most office buildings with high structural vacancy are typical office buildings. Though conversion is a sustainable way of coping with structural vacancy, the financial feasibility or, rather, the financial revenues of conversion as opposed to other possible strategies decide the conversion potential of these office buildings. The financially most interesting strategy can be found by calculating the net present value (NPV) for different possible strategies. Using market conform calculation methods, the NPV was calculated for a sample of structurally vacant office buildings in Amsterdam (Muller et al. 2009). As many of the actors in the market were sceptical of building conversion, a worst-case scenario was calculated for the conversion building costs, and a best-case scenario (from the investors' perspective) was used for calculating the value of the structurally vacant office buildings.

The study used an internal rate of return (IRR) of 7% for the different strategies, making a comparison of the strategies possible. The building's value was determined using the cap-rate method, calculated as the product of the estimated first-year rental income divided by the gross initial yield (GIY). The estimated risk and future value thereafter decide the choice for a strategy. The NPV for consolidation was calculated using a 10-year operation period for the existing building, anticipating that 50% of the building will be let during the whole period. The NPV calculation for adaptation was conducted assuming that a newly adapted building was more attractive to office users. Therefore, the NPV was calculated anticipating that 75% of the office building would be let during the whole 10-year period. In the case of conversion, a worst-case scenario was assumed, using 1450 €/m^2 as input variable for the building costs. The alternative was calculated assuming purchase of the land and the existing building and selling the whole development after 2–3 years. The last alternative considered demolition and construction of new housing on the same plot.

In 40% of the cases studied, the NPV calculations revealed conversion to be financially viable. As the purchasing price per building was not known but estimated, a positive NPV would give an idea of the space for negotiation with the current owner of the building. The financial feasibility of conversion could be additionally enhanced by extending the building horizontally or vertically or by adding a commercial programme like retail or leisure functions to the ground floor of the building. The possibilities depend on the location and the building. Not all locations are suitable for retail or leisure functions, although most office buildings are already located in locations where the ground floor has a public character. A vertical extension could be possible for a large amount of the existing office buildings as these were more sturdily constructed than the standard apartment buildings, and so most office buildings could be extended vertically with one to two floors. Horizontal extensions could also be interesting, depending on the size of the building plot. While in the city centre most office buildings are built on small plots or even adjoining other buildings, office buildings on the city edges and in office locations are built on larger plots, providing enough space for extensions. On the other hand, parking possibilities and other functional studies were not part of this study and should be considered in subsequent assessments.

Calculating the value of a structurally vacant office building means that future rent is not expected, and its value would be 0 – except for the value of the remaining land lease minus the costs for clearing the site. Keeping this in mind, the value of structurally vacant office buildings should be lower than calculated in this study. For example, if the building is let for 50% and the remaining 50% is structurally vacant, calculating the building's market value by the cap-rate method using a GIY of 7% is rather opportunistic, as the GIY is based on a completely rented out building. A much higher GIY would be more realistic, and in that case the coping strategies conversion and demolition and new construction would become worthy of consideration.

The calculations considering building costs, new rents or sale are calculated on a highly abstracted level. The building costs for office adaptation are retrieved from numbers on standard office developments; the building costs used for the conversion strategy are based on the highest building costs observed in the 14 *ex post* cases. Though the building costs in the *ex post* studies varied between 500 €/m² and 1500 €/m², estimating the building costs for a conversion projects implies a more thorough study of the specific building, and so the costs calculated here were kept high to adhere to a worst-case scenario. Calculating a best-case scenario would conclude conversion to be the optimum strategy for a higher percentage of the structurally vacant office buildings.

The purchasing price of apartments differs according to location, as does the rental price for offices. Subsequently, the conversion potential in some locations was much higher than in other locations. The city centre scored well on conversion potential (76% of the structurally vacant office buildings), together with the location Baarsjes (95%), but also small, old office locations, enclosed within popular housing areas, such as Westerpark, Oud-West, Geuzenveld/Slotermeer and Oud-Zuid show a conversion potential above 70%. On the other hand, some locations with high structural vacancy like Zuidoost, Oost/Watergraafsmeer and Zuideramstel all have a conversion potential lower than 40%. The reason is that all the office buildings in these locations are relatively new, and also the structural vacancy of most of these buildings is only partial, and so the value of the office buildings is relatively high. As the structural vacancy of these buildings would endure or increase, the building's value would decrease, and such depreciation would cause conversion to become the best coping strategy for an even higher percentage of the structurally vacant office buildings. This study shows the differences between the various coping strategies and, most importantly, the importance of the different variables in a development scheme.

6.5   Durability and Sustainability

The Brundtland Commission defines sustainable development as 'development that meets the needs of the present without compromising the ability of future generations to meet their own needs' (WCED 1987). This definition can be broadened to discuss the expected lifespan of a system or of parts or aspects in the system. From this proposition sustainability goals can be defined and achieved, nowadays resulting in climate emission restrictions and aims for reduction. While during the 1990s sustainable development was mostly understood as development without growth, this train of thought has prevailed for the ideas of cradle-to-cradle developments that consider recycling or upcycling of second-hand building materials, meaning the process of converting these materials into new materials or materials of better quality or environmental value (McDonough and Braungart 2002) or extended lifespan (De Jonge 1990) that reduce waste production and energy use in construction.

In the Netherlands, the building industry is responsible for 25% of the road traffic, 35% of the waste produced and 40% of the energy consumption and CO_2 emission (Lichtenberg 2005). Several approaches for reductions were proposed. One example is the proposal for industrial, flexible and demountable (IFD) building systems. IFD buildings are constructed of industrial developed modules that can be disassembled when they are no longer needed or whenever major adaptations are required. The scope of the IFD system is to develop adaptable, transformable buildings that are better suited to accommodate the functions of a society with user preferences that are rapidly changing (Durmisevic 2006). It is meant to decrease the production of waste from the demolition of buildings and at the same time trigger the reuse of building material, thereby limiting the production of new building materials and the emission of CO_2.

Extended building lifespan is another proposal to improve the sustainability of the built environment by increasing the durability of the buildings, considering the building's technical, functional and economic lifespan (De Jonge 1990). From this perspective, change is seen as a constant, while the implications of the changes are seen as uncertain (Leupen 2006). The building is seen as a frame, or bookshelf, and possible functions or activities are seen as books. A slightly different approach focuses on buildings that are robust and flexible in use but specific in appearance and that can accommodate several programmes, like the 'solids' of housing corporation Het Oosten (2004). By developing new buildings with a sturdy structure and without a functional programme, a proposal is made for buildings that should become dear to its users and that should last for at least 100 years.

According to Zijlstra (2006), durability is a product of continuity and changeability. Changes react to continuity, and although demolition of a building is also a change and a possible reaction to continuity, the continuity would then be broken and durability would not be the result. Changes add quality to existing buildings and make new programmes possible. The lifespan of buildings may be prolonged as a result of adaptation, whether considering adaptation of a building for the original function or considering conversion of a building. In housing, if a building is adapted, an expanded lifespan of 30 years following the adaptation is expected (Douglas 2006). Rental offices respond much faster to market and user preferences, and so the extended lifespan expectation of each adaptation will be lower.

In this research, durability is seen as one of the main contributors to sustainable buildings and urban areas. As conversion contributes to a longer lifespan for functionally obsolete buildings, developments by conversion are by definition sustainable.

6.6   Conclusion

Adaptive reuse is a way of dealing with obsolete or structurally vacant buildings. The alternatives are consolidation, adaptation or upgrading, or demolition with possible new construction on the site. Consolidation is often chosen as a solution – to do nothing and wait for better times. Often this solution is chosen because it is the easiest option, and in many cases, property owners do not always have the knowledge that is needed to assess the future value of an adapted building. The several instruments that are presented here may be of use for studying the adaptive reuse potential of buildings and locations. The transformation meter is an instrument that is easy to use for a first quick scan for residential conversion. The programmatic quick scan adds an analysis for defining a new function for the redundant building. The study of architectural value is an interesting instrument to use parallel to a conversion feasibility analysis, while the architects' method is not as such an instrument but rather the methodical way in which architects approach a study of adaptive reuse potential. Finally, an instrument for studying adaptive reuse potential that has parallels to the transformation meter was developed in practice. In this research, the definitions and characteristics described in the transformation meter have been used, as it uses the same terminology as other assessment tools for building obsolescence and adaptation.

In Part III of this book, multi-criteria decision analysis is discussed in the context of its application to the assessment of adaptive reuse potential; choices between consolidation, adaptation and demolition/new build; and the simultaneously 'greening' of existing building stock.

References

Coenen, J. (2007) Transformatie als architectonische opgave, in *Transformatie van kantoorgebouwen thema's, actoren, instrumenten en projecten*, Van der Voordt, T., Geraedts, R., Remøy, H. and Oudijk, C. eds., Rotterdam: Uitgeverij 010.

De Jonge, H. (1990) The philosophy and practice of maintenance and modernisation, in *Property maintenance management and modernisation*, Quah, L.K. and International Council for Building Research Studies and Documentation eds., Singapore: Longman.

Douglas, J. (2006) *Building adaptation* (2nd edition), London: Elsevier.

Durmisevic, E. (2006) Transformable building structures: design for disassembly as a way to introduce sustainable engineering to building design and construction, PhD Thesis, Delft University of Technology, Delft, The Netherlands.

Geraedts, R. and De Vrij, N. (2004) Transformation meter revisited, in *Open Building Implementation*, Kendall, S. ed., Munice, IN: Ball State University.

Geraedts, R.P. and Van der Voordt, D.J.M. (2003) Offices for living in: an instrument for measuring the potential for transforming offices into homes, *Open House International*, 28(3), 80–90.

Geraedts, R.P. and Van der Voordt, D.J.M. (2007) A tool to measure opportunities and risks of converting empty offices into dwellings, in *Sustainable Urban Areas*, Rotterdam: ENHR.

Hek, M., Kamstra, J. and Geraedts, R.P. (2004) *Herbestemmingswijzer herbestemming van bestaand vastgoed*, Delft: Publikatieburo Bouwkunde.

Het Oosten (2004) Solids. Available at http://www.solids.nl. Accessed on 11 August 2013.

Hofmans, F., Schopmeijer, M., Klerkx, M. and Herwijnen, F.v. (2007) ABT-Quickscan, in *Transformatie van kantoorgebouwen thema's, actoren, instrumenten en projecten*, Van der Voordt, T., Geraedts, R., Remøy, H. and Oudijk, C. eds., Rotterdam: Uitgeverij 010.

Leupen, B. (2006) *Frame and generic space*, Rotterdam: 010 Publishers.

Lichtenberg, J. (2005) *Slimbouwen*, Boxtel: AEneas.

McDonough, W. and Braungart, M. (2002) *Cradle to cradle: remaking the way we make things*, New York: North Point Press.

Muller, R., Remøy, H. and Soeter, J.P. (2009) De Amsterdamse Transformatiemarkt: structurele leegstand 4% lager, *Real Estate Research Quarterly*, 8(1), 3.

Oudijk, C.P.A., Remøy, H. and Van der Voordt, D.J.M. (2007) Projectanalyses, in *Transformatie van kantoorgebouwen thema's, actoren, instrumenten en projecten*, Van der Voordt, T., Geraedts, R., Remøy, H. and Oudijk, C. eds., Rotterdam: Uitgeverij 010.

Remøy, H. and Van der Voordt, D.J.M. (2007a) Conversion of office buildings; a cross-case analysis, in *Building Stock Activation 2007*, Kitsutaka, Y. ed., Tokyo: Tokyo Metropolitan University.

Remøy, H. and Van der Voordt, D.J.M. (2007b), A new life: conversion of vacant office buildings into housing, *Facilities*, 25(3–4), 88–103.

Remøy, H., Schalekamp, M. and Hobma, F. (2009) Transformatie van kantoorterreinen. *Real Estate Research Quarterly*, 1(4), 6.

Van der Voordt, D.J.M., Geraedts, R.P., Remøy, H. and Oudijk, C. eds. (2007) *Transformatie van kantoorgebouwen thema's, actoren, instrumenten en projecten*, Rotterdam: Uitgeverij 010.

Vijverberg, G.A.M. (1995) *Huisvestingsbeleid basis voor bouwkundig onderhoud; kantoorgebouwen in eigendom*. Delft: Delftse Universitaire Pers.

Vijverberg, G.A.M. (2001) *Renovatie van kantoorgebouwen literatuurverkenning en enquete-onderzoek opdrachtgevers, ontwikkelaars en architecten*, Delft: Delftse Universitaire Pers Science.

WCED. (1987) *Our common future*, Oxford: Oxford University Press.

Zijlstra, H. (2006) Building construction in the Netherlands 1940–1970: Continuity + changeability = durability, PhD Thesis, Delft University of Technology, Delft, The Netherlands.

7

Examples of Successful Adaptive Reuse

7.1 Introduction

This chapter is based on 15 Dutch case studies of office to housing conversion. The case study evidence includes material from several sources: the situation before conversion was studied through documents, text, photos and drawings and the situation after conversion was studied through documents and visits to the building. Interviews with stakeholders were held to gain insight in the process and to retrieve additional information about the situation before conversion. In any building project, several actors are involved: the client, the developer, the architect, the structural engineer, the installation engineer and finally the contractor. Ideally, two interviews were performed per project, one with the architect and one with the developer. In some cases though, there was no architect involved, while in two cases the client was also interviewed. The interviews were semi-structured, based on an interview protocol (Yin 1989; Mason 1996), evolving during the 6 months' period in which they were held.

In the interviews, project specificities were discussed. The first issue was the project initiative. Typically, the project was initiated by the developer but sometimes by the local municipality or the owner of the vacant building. Following, questions were asked about the spatial programme, the appointed user and feasibility. Also, the relationship with the local municipality and the municipality's role in the project was questioned. Next, questions were asked about the design phase. Usually, information about this part of the project was retrieved from the architect, but due to the project character, other stakeholders played more important roles than they would have done in a typical new construction project. The executive project leader was sometimes the architect, sometimes the developer. Questions were then asked about the construction phase, a stage

Sustainable Building Adaptation: Innovations in Decision-Making, First Edition.
Sara J. Wilkinson, Hilde Remøy and Craig Langston.
© 2014 John Wiley & Sons, Ltd. Published 2014 by John Wiley & Sons, Ltd.

in which technical obstacles typically surfaced. Finally, questions were raised about delivery, use and building management, process evaluation, financial feasibility and user satisfaction.

The buildings were visited and the new situation was photographed. In some cases, the inhabitants were informally interviewed. Photos of the existing situation and the architectural drawings of the building before and after conversion were used. In many cases, these drawings gave a good overview of the existing structure, stairways, elevators and exterior and structural walls, while the interior walls had often been changed. The interiors of office buildings are often adapted without updating or making new drawings. The written documents consisted of magazine and newspaper articles. These were especially useful in the study of buildings that were transformed several years ago and where the interviewees had forgotten important details.

7.2   Dutch Conversion Projects (Office to Residential)

A specific conversion topic is the conversion of structurally vacant office buildings into housing. This study is based upon a cross-case analysis of 15 cases with the purpose of identifying risks and opportunities for conversion. The cases were chosen by random purposeful sampling (Miles and Huberman 1994; Patton 2002) to make it possible to generalise the findings within a specific group of conversion projects (Flyvbjerg 2006): they are all examples of conversion from offices into housing. They are of significant size (the smallest counted 18 apartments). The adaptation cases chosen were carried out between 1999 and 2012, since the legal framework stayed more or less the same during this period. Table 7.1 gives an overview of the 15 cases that were studied. The studies were performed after the conversions were completed.

Table 7.1   Overview of case studies.

Project		Construction	Transformed	Units	Type of dwelling
1	De Stadhouder	1974	2005	70	First homebuyers, buy
2	Lodewijk Estate	1954	1999	24	Seniors, buy, rent
3	De Enk	1956	2006	69	First homebuyers, buy
4	Schuttersveld	1915–1923	2003	104	Luxury, buy
5	Westplantsoen	1970–1980	1999	45	Students, rent
6	Billiton	1938	2004	28	Luxury, buy
7	Hof ter Hage	1935–1967	1998	97*	Mixed, buy
8	Wilhelmina Estate	1969	2007	43*	Mixed, buy
9	Granida	1958	2005	30*	Luxury, rent
10	Residence De Deel	1959	1999	18	Seniors, buy
11	Twentec Building	1960–1965	2002	87*	Luxury, rent
12	Eendrachtskade	1980	2004	83	Students, rent
13	Churchill Towers	1970	1999	120	Mixed, rent
14	Puntegale	1940–1946	1999	210*	First homebuyers, rent
15	Westerlaan Tower	1966	2012	45	Luxury, buy

*Other functions were added, such as small-scale offices, shops, healthcare and commercial space.

The original buildings represent three different construction periods: buildings constructed before 1950, buildings constructed between 1950 and 1965 and buildings constructed between 1965 and 1980. The typological characteristics of office buildings changed over time. Buildings from 1950 to 1990 have the highest proportions of structural vacancy, while office buildings built after 1990 are more popular. These findings correspond with the obsolescence of the supply from this period (Remøy 2007). The vacancy concentrates in the buildings from 1970 to 1990, while in buildings from 1950 to 1970, it is less significant. An analysis is performed concentrating on the office buildings developed between 1970 and 1990, revealing their typological characteristics.

7.2.1   'Stadhouder' in Alphen aan den Rijn

Completion office building: 1974
Completion conversion: 2005
Commissioner: Giesbers Maasdijken Ontwikkeling
Architectural design conversion: Herms van den Berg
General contractor: Giesbers Maasdijken Bouw
Area: 5500 m²
Area after conversion: 7500 m²

In 2002, the owner of the office building foresaw vacancy and contacted the property developer to make a plan for conversion of the building or the site. Within the zoning plan the site was designated for housing. The potential profit was calculated for both conversion and demolition and construction of a new housing development. The revenues for conversion were the highest. The target group for the new development was starters in the housing market. Commercial developments of the ground level were considered; however, a market analysis was not positive, and so the base was designated for entrances and storage space for the housing. The structure of the building turned was robust, and two extra floors could be added on top of the building. By adding maisonettes to this part of the building, the existing staircases and elevators could be reused. The facade was technically outdated and was removed. The floors were made of prestressed concrete slabs that did not allow for holes for large vertical shafts. The service shafts were therefore placed on the outside of the building's façade. The contractor, a subdivision of the developer, was involved in the project from the beginning, together with the architect. This and strict adherence to the zoning plan enabled a short development and construction period: from the first sketch it took 2 years to complete the conversion.

7.2.2   'Lodewijk Staete' in Appingedam

Completion office building: 1954
Completion transformation: 2002
Commissioner: Woongroep Marenland (housing association)

Architectural design conversion: Martini Architekten
General contractor: Bouwbedrijf Kooi
Area: 4400 m²

The 'Willem Lodewijk van Nassau Kazerne' was built in 1954 as quarters for the air force, after which it was used as governmental offices until 1999. Even before the government moved out, a conversion feasibility study was performed by the owner, the Dutch Government (RVOB). The Marenland housing association bought the building to convert it into apartments for elderly people. Twenty-four apartments, partly privately owned, partly for rent, were developed. Designing for the elderly implied spacious apartments that could accommodate a wheelchair. The architect calculated the construction costs, by comparing the project to earlier conversions. The large entrance on the ground floor was reused and connected to a new atrium at the back of the building with entrance galleries, an elevator and staircases making the second and third floor accessible for wheelchairs. To the façades, balconies and sun porches were added, and on the third floor these were cut out of the roof. The main structure of the third floor diverged from the first and second floor and also the floors were thinner, so no extra weight could be added to the third floor. For the same reason the acoustic insulation was insufficient, and floating floors and suspended ceilings were added.

The large entrance hall is appreciated by the residents, a quality quite specific to the converted buildings. The building is also widely appreciated for its historical meaning and value.

7.2.3   'Enka' in Arnhem

Completion office building: 1956
Completion conversion: 2006
Commissioner: BAM, Klaassen
Architectural design conversion: Bureau voor Harmonische Architectuur
General contractor: BAM and Klaassen
Area: 6600 m²

The building was developed as the headquarters of Akzo Nobel. By the end of the 1990s, it became functionally obsolete and was sold. The building is a listed monument and may not be demolished. Given the advantageous housing market, the owners chose to transform the building into housing for first homebuyers. The market pressure in this category was high, and the building was suitable for this purpose: the entrance to the apartments is via a gallery, and since the building is a monument, no balconies could be added. Both characteristics are not favoured by higher-income target groups, while for first homebuyers they are not a problem. The existing staircases and elevators were reused, keeping the character of the building and also adding to the financial feasibility of the project. The building's specific monumental characteristics determined the division of the building into apartments, entrances and public space. Open spaces were kept where artwork had been incorporated. The load-bearing structure turned out to differ between the floors; hence, new inner walls had to be lightweight. Some parts of the ceiling were too low according to the Dutch housing standards, but exemptions from the decree were obtained. The conversion of this monument implied that compromises had to be made regarding both the quality of the apartments and monument and building regulations. The building is well known in Arnhem as the head office of a firm where many people worked. The building still has this identity and imposes it to its surrounding; the whole area is now being redeveloped as the Akzo Housing Estate.

7.2.4   'Schuttersveld' in Delft

Completion office building: 1923
Completion conversion: 2003
Commissioner: ABB
Architectural design conversion: Feekes & Colijn
General contractor: ABB Construction
Area: 16,100 m^2
Area after conversion: 26,150 m^2

The building, owned by the TU Delft, comprised accommodated offices and the university's library. As a new library was built on the university campus, the building became redundant and was sold, however claiming that it would be reused since parts of it were listed. ABB bought the building for conversion into housing, reusing as much of it as possible. Only a small part that was of poor technical quality was demolished. While 10,000 extra square metres were realised, the weight of the new construction had to be kept low for the existing fundaments to suffice. During the design phase, the building was mothballed. Since it was not heated, damage occurred to the heritage character stairwell interior and the entrance space. The storey heights were partly too low according to the building decree, and though exemptions were made, potential buyers dropped out for this reason. The apartments on the ground floor have gardens, and the apartments on the top floor share a terrace on the roof of the new part of the building. The apartments in the new part all have balconies. Balconies for the roof apartments were cut out of the roof. New windows resembling the originals were placed in the original openings. Windows with original glass paintings were kept and insulated by a second glass layer on the inside. After conversion, the whole building except for the new built part was listed.

7.2.5   'Westplantsoen' in Delft

Completion office building: 1970s
Completion conversion: 1999
Commissioner: housing association DUWO
Architectural design conversion: Karina Benraad Architecture office

General contractor: ERA
Area: 5400 m²
Area after conversion: 6000 m²

The former tax office had been vacant for 5 years when the housing association DUWO bought it for conversion into student housing. Demolition and new construction was considered but was regarded to give lower revenues. Based on market studies, DUWO decided to develop two-room apartments, since these could also be let to other target groups. Financial feasibility studies were negative; hence, subsidies were granted by the TU Delft and the municipality. The structure of the building was adaptable with columns and a facade grid of 1.8 m. The façade was technically outdated and looked worn out. Therefore, it was insulated and a new outer layer was added that changed its appearance. The fire escapes were reused, but the main staircase was relocated from the centre to the perimeter of the central axis where a corridor now facilitates the apartment's entrances. The original entrance was situated above street level and could only be reached by outdoor stairs. To allow for wheelchair access, it was moved to the basement. As private outdoor spaces were required by the building decree, balconies were added, providing the building with a completely new appearance. The simple shape of the building contributed to an efficient conversion design. The adaptability of the structure was one of the critical success factors of this conversion, together with the system of floors: these could easily be removed and service shafts could be added.

7.2.6   'Wilhelminastaete' in Diemen

Completion office building: 1969
Completion conversion: 2007
Commissioner: Rabo
Architectural design conversion: Rappange & Partners
General contractor: Heddes Constructors
Area: 6700 m²
Area after conversion: 8500 m²

The former Rabobank office was technically and functionally outdated and finally redundant. The location was no longer considered suitable for offices, and due to the technical state of the building, a retrofit would imply a thorough adaptation. Rabo development took initiative for conversion. Studying the feasibility of conversion, they saw housing for the elderly as an opportunity because of high local demand for housing within this target group. The municipality was supporting the conversion plans. Together with the architect, a functional programme was developed that would fit the building and the intended target group. The building was equipped with spacious entrances to fit wheelchairs and strollers. On the ground floor, a small bank and a library were realised, together with some parking. Due to the technical state of the facade (the sheet stone façade was literally falling off), it was stripped, and a new façade was added. The new façade was designed as a typical housing façade, vertically laid out instead of horizontally as was the existing one. Conversion of the building into housing was feasible because of an adaptive structure of columns with suitable dimensions. The existing elevators could be reused. The existing stairways were reused as escape routes, and no additional staircases were needed.

7.2.7   'Granida' in Eindhoven

Completion office building: 1958
Completion conversion: 2005
Commissioner: Van Straten, Woonveste
Architectural design conversion: Architecture Office Ton Kandelaars
General contractor: Van Straten
Area: 7800 m^2

This office building was built for the municipal health service that moved out in 1995 after which the building served as temporary offices for the municipality. The building was owned by the municipality. They decided the building should not be demolished but rather would be sold to the developer with the best conversion plan. The winning development scheme considered conversion into apartments. The size of the apartments was dictated by the building structure, and large luxury apartments were developed and later bought by the investor Vesteda. The existing staircases, elevators and entrances were reused, resulting in much larger common space than in new apartment buildings. To eliminate the effect of thermal bridges in the structure, the apartments were designed like boxes that fit into the existing structure. The demands of the building decree for new housing were followed, though the elevator doors were too low, the steps of the stairs were too high and in some apartment the floor height is in some places too low. Exemptions were made for these issues. During the construction the measurements and materials of the structure turned out to differ from floor to floor, corrosion of steel reinforcement was discovered, and parts of the structure were deemed too weak, according to the fire-safety legislation. As a result, the 'apartment boxes' that were prefabricated did not fit and had to be refitted, accumulating extra costs. Though the original building was not listed, the municipality of Eindhoven found this building significant for a historical period of architecture related to one of the most important periods of growth that Eindhoven experienced and therefore decided to preserve the building.

7.2.8   'Residentie de Deel' in Emmeloord

Completion office building: 1959
Completion conversion: 1999
Commissioner: WEN

Architectural design conversion: G. Stuwe and C.P. van den Bliek
General contractor: Haase
Area: 1980 m²

The office building was owned by the Dutch Government and accommodated offices for the Water Management Bureau. Its typology with small spaces around a large typing hall made it functionally obsolescent for modern offices. The façade was technically and thermally outdated. WEN saw possibilities for residential conversion and bought it. The typology and structure was seen as suitable for senior citizen housing; the typing hall was transformed into an entrance atrium. Though the dimensions of the structure were unsuited, parking could be realised in the basement. The floors, an early example of precast concrete floors in the Netherlands, were thin and did not have sufficient acoustic insulation to meet housing standards. To improve the acoustics and meet the new standards, an extra layer of concrete was added on top of the existing floors as well as new suspended ceilings below. Because of the heavy floors, the walls had to be light and were built from steel frames and gypsum boards. Originally, the building had a curtain wall façade, a light façade hanging on the outside of the construction. To spare weight in the new design, lightweight concrete was applied. Balconies were designed as loggias. All parts of the building turned out to be in a poorer physical condition than estimated, accumulating extra building costs.

7.2.9   'Twentec' in Enschede

Completion office building: 1960–1965
Completion conversion: 2002
Commissioner: Dura TePas
Architectural design conversion: A12 architects
General contractor: Dura TePas
Area: 11,940 m²

Originally, the two towers on a single base were developed as offices for the textile industry. From 1977 onwards the buildings were used as offices until becoming redundant in 1995. The municipality was planning to redevelop the area, including a large parking garage partly under one of the towers. To boost new developments, Vesteda, the owner of the towers, decided to reuse one tower while demolishing the other. Because of estimated low building costs and short development time, conversion was favoured over demolition and reconstruction. Luxury apartments for senior citizens were developed. The existing façade was technically and visually outdated and was removed and replaced by a new façade. The dimensions of the existing structure were appropriate for conversion into apartments. Since the floors were cantilevered, balconies could not be added to the facade. However, private outdoor space was required, so enclosed loggias were placed. Enclosed loggias implied a fire-safety problem, imposing the use of highly fire-preventing materials, raising the building costs. The estimated short development period was a motive for conversion, but eventually the conversion took longer than planned, causing rental losses and higher conversion costs. However, when finally completed, the building's commercial success led the actors involved to reflect favourably about the conversion decision.

7.2.10   'Eendrachtskade' in Groningen

Completion office building: 1980
Completion conversion: 2004
Commissioner: housing association Stichting In
Architectural design conversion: Scheffer van der Wal/Stichting In
General contractor: Ballast Nedam
Area: 3800 m^2

This office building was property of the ING and was vacant from 2002. The housing association found the building had potential for conversion into student housing. Initially the purchase price was too high, but as the general contractor agreed to be part of the development team, the building was bought. The housing association owned other student housing units in the area and found the property good for increasing the quantity housing in the area. To keep the building costs low, the housing association sought to reuse as much as possible of the original building. The building 'was not a beauty' but quite new, with a technically well-functioning façade. Each bay of the structure provided at least one window that could be opened, so the facade was unchanged. The ground floor and the entrance were changed slightly to give space for postboxes and doorbells for the 84 individual studios. The stairways and the elevator were reused. The studios could fit efficiently into the existing structure. Placing vertical shafts for services in the prestressed concrete floor was problematic; the reinforcement bars had to be located using a detector. The time span from purchase to delivery of the studios was 1 year and according to planning. As the building was relatively new, there were no unpleasant surprises during the construction period, and the final building costs were close to the cost estimates.

7.2.11   'Billiton' in Den Haag

Completion office building: 1938
Completion conversion: 2004
Commissioner: Van Hoogevest
Architectural design conversion: Van Ede Architecten
General contractor: Van Hoogevest
Area: 25,000 m^2

The office building was built for Billiton in 1938. Billiton became part of Shell, and the building was vacated and sold to an investor in 1988. However, the building was functionally outdated and partly vacant, partly rented out to a school for low rents. Van Hoogevest saw conversion potential in this building and contacted the architect who undertook a feasibility study of the possibilities of housing, with positive results. The building was technically in good state, despite some corrosion of steel reinforcement in concrete elements. Demolition was not an option since the building was a listed monument. The zoning plan allowed for housing and other reasons for conversion into housing were the image of the building, the measurements and adaptability of the structure and good parking possibilities in the basement. An approximate programme of requirements was made by van Hoogevest, although the building was the decisive factor. As the building was a listed monument, the front facade could not be altered. On the rear elevation, windows were enlarged and balconies were added. Two extra stairways with elevators were added. High groundwater level had to be taken care of to make the basement suited for parking. By using the 'box-in-box' principle, the acoustic insulation between apartments was improved. The box-in-box principle means that by adding floating floors and suspended ceilings, a room or apartment is effectively insulated without adding much weight to the structure. The insulation of the façade was improved by adding insulation on the inside. The Monument Act determined the level of intervention in the building. However, fire-safety regulations had to be met, and also the level of comfort described in the building decree was mandatory.

7.2.12   'Hof ter Hage' in Den Haag

Completion office building: 1935/1967
Completion conversion: 1998
Commissioner: BAM
Architectural design conversion: ONB, Witt & Jongen
General contractor: BAM
Area: 22,000 m²
Area after conversion: 25,000 m²

This city block of offices was built over 30 years for the National Mail and Telecom Company. As the telephone was digitalised, the building become functionally obsolete and was sold to BAM, who saw the building as a prestigious housing project on a prime location in Den Haag. Financial and functional feasibility studies were conducted, concluding to convert parts of the block and to settle for demolition and new construction for less adaptable parts. A short programme of requirements was set up, locating shops on street level, while the upper floors would be converted into housing. The basement was assigned for parking. A design and build contract was applied, involving all parties in an early stage of the project. The existing building was dimensioned for heavy floor loads and was easily adapted. Balconies were located on the courtyard side since the streets are quite noisy and to maintain the building's appearance. Though the building was not a monument, it had character and renown. A variety of apartments were developed, depending of the structure and street facade of that part of the building. Large apartments accessible by a stairway and elevator per two apartments were located on the widest streets. The conversion process was complicated because of its inner-city location: there was no construction site area. Also, the building was very robust, which made the partial demolition challenging and a reason that the final construction costs were higher than estimated.

7.2.13   'Churchill Towers' in Rijswijk

Completion office building: 1970
Completion conversion: 1999
Commissioner: Geerlings Vastgoed
Architectural design conversion: Oving Architekten
General contractor: Gebroeders Verschoor
Area: 20,700 m^2
Area after conversion: 24,000 m^2

The two towers were built in 1970 on a common base accommodating parking. In the beginning of the 1990s, the building became redundant. It was seen as functionally outdated with deep floors intended for open office landscapes. The building was in technically good state, though the façade was outdated. A design and build contract was used for the conversion. The architect and the developer together studied the functional and financial feasibility for conversion into housing and decided to let the buyers decide the size of their apartments. A model apartment was built to give potential buyers an idea of the possibilities. The thermal insulation of the façade was insufficient for housing. The original glass façade was kept, and a second façade was placed on the inside. In between these layers, private outdoor spaces were realised. The original concrete corners, constructed to enhance building stability, turned out to be superfluous and were removed, opening up the façade. The existing stairways and two elevators were reused. Two elevators were removed, and the shafts were reused as service ducts.

7.2.14   'Puntegale' in Rotterdam

Completion office building: 1940–1948
Completion conversion: 1999
Commissioner: housing association Stadswonen
Architectural design conversion: De Jong Bokstijn Architecten
General contractor: Moeskops' bouwbedrijf
Area: 25,700 m^2

Built as offices for the tax and customs administration, the building became redundant when this service moved to a new building in the beginning of the 1990s. The building was nominated for listing, and could not be demolished, but was functionally obsolete. The sturdy structure, large dimensions and character of the façade made it potentially attractive for housing. Stadswonen had experience with conversion and found this building interesting for conversion into housing for students and first homebuyers in the housing market, the so-called young professionals. Though thermal and acoustic insulation was applied on the internal face of the façade, the building is situated on a heavily trafficked road, and acoustically some parts of the building were unsuitable for housing. These zones, along with the ground floor, were assigned for small offices. The building had high floors, and the three lower floors were so high that maisonettes could be provided. The original entrance and stairways were reused. New elevators and fire escapes were added. The original paternoster elevator was kept though not for daily use. The apartments were made accessible by corridors. The rear of the building had galleries that were reused for private outdoor spaces for the apartments on this side of the building. Since the building is a monument, balconies could not be added, and the apartments on the front side have no outdoor space. Instead, a common roof terrace was provided. Stadswonen sees the building as a trigger for new development in this area. The conversion is financially viable in the long run, since the value of the building and the location are expected to increase.

7.2.15   Westerlaan Tower in Rotterdam

Completion office building: 1966
Completion conversion: 2012
Commissioner: Calandlaan CV
Architectural design conversion: Ector Hoogstad Architecten
General contractor: Dura Vermeer
Area: 8500 m² office space, 45 apartments (7800 m²)

The office tower was constructed as a group of buildings, including a low-rise building. The site accommodated the headquarters of Vopak, a multinational oil-related company. The building became functionally obsolete and redundant in 2002, as Vopak had the low rise adapted to fit new functional demands as less space was needed. The location of the building, next to a park with a view on the river and in a quiet housing and headquarters area, made it possible to redevelop the tower as a functional mix of offices and high-end housing. The upper floors (11–19) accommodate five apartments per floor, varying from 95 to 195 m². The sturdy structure and façade and the reuse of stairs and elevator shafts made conversion possible. The top four floors were demolished, and six new floors were added. A new entrance was formed to connect the tower to a new underground parking garage. Apartments and offices were reached by a central corridor around the elevator shafts and emergency staircases. All the apartments had large balconies added to the façade. Since the balconies had to be supported by the existing construction, they were built in lightweight material and hung from the façade. The development was supported by the municipality and is part of the municipal policy of stimulating city centre living.

7.3   Discussion

7.3.1   Data Analysis

For each project, project and process descriptions were written, distilled from the interviews and the written documents. The drawings and photos were used to explain each case. The case studies were written based on the interviews and sent to the interviewees for feedback and confirmation of accuracy. When the interviewees did not agree with the account, a second round of consultation was held. Case studies that were written up from only one interview were validated by another stakeholder in the same project. A cross-case analysis by manual comparison was the last step. The data were arranged in a matrix and analysed for patterns (Yin 1989). As a result, the projects could be divided in three categories: buildings from before 1950 (or designed before 1950), buildings from 1950 to 1965 and buildings from 1965 to 1980.

The five buildings constructed before 1950 share several characteristics; they are monumental in their appearance, and three are listed monuments. The buildings have structural, solid outer walls and considerable size. Four were built to accommodate specific governmental services. Five buildings were built between 1950 and 1965. During the 1950s, new construction methods entered the market. The buildings of the Akzo headquarters, De Enk, in Arnhem and the GGD Building, Granida, in Eindhoven have structural columns in the facade with additional columns in the centre of the building, while the Estate De Deel and the Twentec Building are early examples of structures with columns and open floor plates. Of the five buildings newer than 1965, one of them has a structural facade in the form of facade columns, and four have a construction of columns and open floor plates. Of these four, in two of them the columns are placed directly behind the facade, while in the other two the facade is kept completely free from the construction.

7.3.2   Conversion Risks

Former studies (Geraedts and De Vrij 2004; Geraedts and Van der Voordt 2007) developed checklists to determine development risks and finally instruments to decide office buildings potential for conversion. The cross-case analysis of the 15 cases was performed focusing on the risks and unforeseen problems that surfaced during the building phase of the project. The four risk categories (legal, financial, technical and functional/architectonic) of these studies were used.

The projects were all completed, which implies that the requirements made in the zoning law and in the building law were met satisfactorily. Asbestos was found in 7 of the 14 projects. The removal of asbestos follows strict rules and therefore incurs high expenses. In all the projects, the eventual removal of asbestos would be paid for by the seller of the building. What was stated as a risk in previous research is taken into account by developers of conversion projects and has gone from being a risk to being a cost that can be calculated.

Apartments in the projects studied were let or sold without problems, except in a few cases; in one case luxury apartments without a private outdoor space and with incidentally low ceilings (not according to the building rules) were sold only after lowering the price. In another case, some apartments with daylight from the north only were not sold for the initial asking price. In all cases, the difficulty of selling the apartments in conversion projects did not exceed that in other projects. In some cases, model apartments were furnished to boost the sale, occasionally even before initiating the conversion. Any developer, assisted by an architect, needs to be aware of the users' wishes. Even in a tight housing market, quality and willingness to pay correspond, especially in the upper part of the housing market.

Three out of five buildings from before 1950 were not built according to the construction drawings, or the construction differed and had different measurement from floor to floor. In one of the five projects, the differences were anticipated from the start; the floors were radically different. In the first years after World War II, housing was prioritised over commercial in the Netherlands. It was difficult to get building materials, and in many cases contractors used the materials they could find without changing the drawings. Two of the five buildings from 1950 to 1965 were not built according to drawings, and the construction materials and measurements were different per floor. The buildings constructed after 1965 showed no such differences. The risk of inaccurate drawings and differing construction is strong in buildings dating from before 1965. Building methods and measuring methods were not very precise.

The main structure was found to be in an unsatisfactorily state only in 1 of the 15 projects: Granida. Since the concrete of the external columns contained corroded steel, it had to be renovated and reinforced. The repair itself added extra costs to the project, but additionally, as a result of the repairs, the columns became wider, and the design needed to be modified. In another project, Billiton, concrete reinforcement corrosion was found in part of the façade but required only a minor investment to repair. Rotting timber, corroding reinforcement in concrete or oxidising steel can increase conversion costs, but in most cases these problems are visible in the preliminary survey and appraisal and will not be a high risk. Providing additional load-bearing capacity to the structure was problematic in one case only. Office buildings are constructed to carry more loads than housing; hence, in most cases an additional floor can be carried by the existing structure.

Apartments are normally smaller than office units, and more shafts are needed for electricity, water and plumbing. In the buildings constructed before 1965, floors were penetrated and shafts were placed without problems. After 1965, reinforced concrete was commonly used, making larger span without columns possible. The problem with reinforced concrete is that it loses strength when the reinforcing steel bars are cut. In three of the five buildings built after 1965, reinforced concrete was used. Nowadays, reinforced concrete is the most common structural material used in buildings. When adapting or converting a building constructed before 1965, the construction method should be taken into account. Designing apartments with a minimum of shafts is a challenge for the architect. The problem can be solved; the accurate place of the steel trusses can be located with accurate metal detectors, like it was in the Eendracht project.

Reinforced concrete was not used in Dutch building constructions before 1965. The measurements of the structural grid were smaller, and the small spans came with thinner, lighter floor slabs. These types of floors are strong; they are constructed to allow for the loadings of office equipment, which before 1965 were heavier than now. The problem of converting these structures into housing is the acoustic insulation of the floors. It is reasonable to conclude that floors constructed before 1965 need acoustic upgrades improved to meet the requirements of modern building standards. This can be done, as seen in the cases, by adding a floating floor and a suspended ceiling.

The Dutch building code requires a higher level of thermal and acoustic insulation of the façade for housing than for offices. The façades in six of the buildings were removed and new façades were added. In seven projects, the thermal and acoustic insulation of the façades was improved; for five of the projects, there was no other possibility because these were monumental. The façade of one project only was not altered.

In the initial phase of a conversion project, before deciding to buy the property, the developer, alone or with the architect and other experts, made quick scans of the possibilities for conversion. Sketches were made, when possible based on the original drawings, to make an estimation of the possibilities to fill in apartments or other functions.

In addition to the risks already identified, some new risks appeared in this study. The municipality not allowing exceptions from the zoning plan is a risk. However, based on these cases, the risk was recognised of the municipality slowing the process where a change or an exception of the zoning plan was needed. One of the advantages of conversion projects is the short time span from first sketch till delivering the apartments. Time-consuming procedures may slow the process and delay income, making projects financially unviable.

When a first scan is made of the building to adapt, the height of the floors needs attention. In most cases, office buildings have higher floors than required for apartments, but when both a floating floor and a suspended ceiling are needed, additional height is required. To be sure of sufficient floor height, a clearance of 2.60 m is required inside the apartments, and the floor-to-floor height should be 3 m, allowing for mechanical ventilation above the suspended ceiling and a minimum height of 100 mm for the floating floor.

The financial feasibility can be affected by other risks. While a lowered ceiling and floating floor can be installed, constructions repaired and reinforced, shafts cut through reinforced concrete floors and municipalities asked to make changes or exceptions to the zoning plan, the conversion costs will increase as a result. Several of the risks recognised by Geraedts and De Vrij (2004) could easily be assessed in the initial phase of the conversion and are not seen as problems. After analysing the 15 cases, the risk list could be shortened (Table 7.2).

Table 7.2 Risks identified from the case studies.

Aspect	Risks
Legal	*Zoning law*: impossible to meet requirements (function, form, size) *Dutch building decree*: impossible to meet requirements from the VROM (2003), including noise-level prescriptions and fire precautions *Municipal building act*: the municipality is unwilling to cooperate
Financial	*Development costs*: slow handling of procedures (loss of income) *Vacancy*: failing incomes from exploitation or sale of the property
Technical	Incorrect or incomplete building structure assessment Inadequate/poor state of the main structure or foundation (rotten concrete or wood, corroded steel) Insufficient shafts available; construction allows no extra penetrations or shafts being made Inadequate acoustic insulation of the floors/thin floors Insufficient thermal and acoustic insulation in the facades Insufficient daylight for housing
Functional/architectonic	*Incorrect assessment of functional possibilities*: preliminary sketches prove worthless; 'unusable' space

7.3.3   Conversion Opportunities

The short development time span from the first sketch till delivery of the apartments is an advantage for conversion projects. The project Stadhouder was developed in only 2 years from the first sketch till delivery. While still working on the design, the facade was removed and the building stripped down to structure, stairs and elevator. Not only was time saved because the main structure was already there, but also because of this, there were fewer working days lost due to bad weather.

The 'WYSIWYG factor' is another advantage for conversion projects: What You See Is What You Get. Model apartments can be furnished already before demolition starts. Most people cannot interpret architectural draw-ings, while this communication form may better inform potential buyers and boost the apartment sales. The project De Stadhouder demonstrates that the financial feasibility of conversion projects can be improved by taking advantage of fiscal rules according to geographical location. An example from the Netherlands increasing the financial feasibility of conver-sion projects is the reduced conveyance duty on land and building (6% instead of 19%). If the apartments are sold within 6 months, the conveyance duty only has to be paid for by the buyer of the second sale. The full VAT (19%) is then only paid for the building activities.

Conversion of vacant offices is an opportunity for development in an already organised context, in central urban areas. The conversion of an already existing building normally attracts fewer objections from neigh-bours or neighbouring users than the demolition of an existing building and new construction. Finally, the redevelopment of a building in an area of high vacancy, obsolescence and dilapidation can give a boost to the area and increase the value of the land within reasonable investment time perspectives. This gives developers and investors a chance to increase the financial feasibility of a project, for both social housing corporations and corporations active in the unregulated housing market.

Finally, conversion of vacant offices is a sustainable alternative to demolition and new build. Transforming vacant office buildings into housing uses embodied carbon, saves building materials and building material transportation and produces less waste than demolition and new construction. A frequently heard argument for demolition is that the thermal insulation in older buildings is not adequate. Demolition is in this case used as a sustainability argument. However, the case studies show that the performance of existing office buildings can be adapted to the level of the Dutch building legislation law as well as to the level of comfort expected by the relevant user group.

Most of the risks that were recognised in the cross-case analysis were found in the technical category. The risks within this category turned out to depend on the date of construction of the existing building. Fewer technical risks were experienced in the conversion of the five buildings that were originally constructed between 1965 and 1980. The construction drawings of these buildings were correct, and the building condition was good. The floors in the buildings from the later part of this period, De Stadhouder, Westplantsoen and Eendrachtskade, had sufficient acoustic insulation for housing. The Eendrachtskade had double glazing. The thermal insulation of the facade was sufficient for housing, but the acoustic insulation was not. In this case, the municipality made an exception from the building code since students (the expected occupants of the transformed building) are considered to tolerate noise well. If the relevant user groups had been seniors, the acoustic insulation of the façade would have required improvement in all likelihood. In the financial category few risks were recognised. However, all technical, legal and functional risks reduce the financial feasibility of the project. Hence, it may be concluded that most risks are also financial risks.

The projects included in this chapter are completed conversion projects. One of the legal risks was the municipalities' cooperation on zoning plan changes and building code exemptions. However, the parties involved in all 15 conversions were satisfied with the municipalities' cooperation. One question that remains unanswered is whether the projects would have failed without municipal cooperation.

In the analysis of the 15 cases, the aim was to reveal the factors that influence the projects' financial feasibility. The developers who were interviewed stated that the earnings on conversion projects are too low compared to new construction. Also, other actors in the conversion processes complained about overrun budgets and too many hours spent to develop specific solutions to problems that occurred during the building process. Of the 15 cases, only one developer was willing to share financial information regarding the project. In one case, the developer informed us that the financial goal was not achieved because of insufficient sales. In the other 14 cases, the developers claimed that there were no financial losses, despite the fact that the budgets were overrun.

Table 7.3 Typological characteristics that affect building conversion potential.

	Positive	Negative
Structure and floors	Structural grid 5.4 m or 7.2 m: common in housing Columns; free plans Constructed for heavy carriage; 2.94 kN/m² required, 1.72 kN/m² required for housing	Dense grids Low ceilings under existing beams *Reinforced concrete*: complicates floor penetration *Thin floors*: acoustic insulation insufficient
Facade	Facade grid 1.8 m and load-bearing walls: possible to attach interior walls to facade	*Curtain walls*: inadequate technical state, no attachment points for interior walls *Cantilevering floors*: complicates adding balconies Remove and rebuild facades
Floor layout, length and depth	Depth of buildings normally enhances the transformation potential	Centrally placed elevators and staircases
Stairs and elevators	Excess number of elevators	Insufficient number of escape routes Space occupied by stairs and elevators: excessive

7.3.4   Typology

The lessons learnt from the *ex post* case analysis show how opportunities and obstacles of conversion are closely related to the architectural characteristics of the existing buildings. To be able to use the information from the *ex post* cases to study other buildings *ex ante*, the conversion potential of the common office types was studied theoretically. This study concerned finding the characteristics that describe a building type and that also influence its conversion potential, for example, structure and floor span, façade characteristics, floor layout and the length and depth of the building and the number and situation of stairs and elevators. These characteristics were identified through case studies (Remøy and Van der Voordt 2007a) and are summarised in Table 7.3.

7.3.5   Structure and Floors

The main load-bearing structure in standard office buildings most commonly has a span or bay width of 7.2 m, though a structural grid of 5.4 m is sometimes used and in some cases a larger grid of 8.1 m. In Dutch housing 5.4 m has long been a standard measurement for the width of single-family housing and apartments, while in some newer apartments a structural grid of 7.2 m or even 8.1 m is used. In conversions of older office buildings with flat-slab beamless floors, the large number of columns may cause a partition problem. The linear structures from the 1980s have larger spans perpendicular to the long walls and are more easily adapted to new use. Beams under the floors may cause problems, because the free height of the floors is lowered, and when adapting new installations, these need to be fitted in under the beams. The floors in office buildings are normally constructed to carry more weight than in housing (in offices, 2.94 kN/m^2 is required; in housing, 1.72 kN/m^2), a positive characteristic for conversion of these buildings. However, office buildings are normally constructed using precast concrete floors. The limited possibility of penetrating these floors makes it difficult to add vertical shafts. The reinforcement bars within the floor slabs may be located, but are not always are they in the same place on all floors. Floors of this kind may be of hindrance for conversion, but do not make it impossible. Solutions could be found by smart reuse of the central existing shafts.

The floor height of office buildings is normally sufficient for conversion to housing; the Dutch building decree requires 2.6 m free floor height in housing. The acoustic insulation of typical office building floors is not sufficient for housing. In most conversions, adherence to the building code requires the addition of acoustic insulation following the box-in-box principle, with a suspended ceiling and a floating floor. When lowering the floor height locally because of existing beams, exceptions from the requirements to the free floor height may be given.

7.3.6   Floor Layout, Building Length and Depth

An efficient layout of housing on an office floor may be thwarted by the office floor plan. The location of the central elevator and staircases may be inconvenient for housing. Moving the elevator core and staircase is usually not possible, because the core also contributes to the stability of the structure. Placing a new elevator in many cases would only be possible outside the existing building, so that no extra shafts have to be made in the existing floor. However, applying radical changes to the building's staircases or elevator cores critically increases the building costs of conversion.

The depth of office buildings is mentioned as an obstacle for residential conversion, but the normal depths of Dutch office buildings built in the 1980s are actually similar to the normal depths of Dutch apartments. In many cases, the depth of office buildings is even a positive aspect when considering conversion. The building depth may be an obstacle for the conversion of older office buildings; buildings from the 1960s were generally deeper and with less daylight access.

7.3.7   Façade

Interventions in the façade represent the most substantial costs and represent a critical factor for the financial feasibility of conversion projects. Completely replacing the existing façade of an office building implies a high conversion building cost, though it was found necessary in 7 of the 15 cases that were studied. In examples of conversion of office buildings into student housing, the financial targets were met because the façade could be retained. When office buildings are converted into more expensive housing, more substantial changes may be undertaken.

The Dutch building code was recently altered. Until 2003, a balcony or other private outdoor space was mandatory. Though for a few years the new building decree did not prescribe a private outdoor space, most people demand or desire dwellings to have a balcony or a terrace, especially in expensive high-end apartments, and so the decree was changed back in 2008. Until the 1970s, modernism had a great effect on the design of floor plans. The Domino principle by Le Corbusier was incorporated in its pure form with columns and floors, with cantilevering floors and a curtain wall facade. Such a structure though makes the addition of balconies difficult. Adding a loggia is an alternative; the floors need to be wrapped in insulation, and an undesired height difference between inside and outside would be the result. French balconies or winter gardens are possible solutions.

7.3.8   Stairs and Elevators

Office buildings are designed for more people per square metre and more traffic than apartment buildings. Therefore, the number of elevators available for the new function is a positive aspect for conversion. Elevator shafts that are not needed after the conversion into housing may be reused as shafts for ventilation, electricity, water supplies and sewer. Since the shafts are often used to provide for the stability of the structure, the possibility for alterations or making holes in the shaft walls may be restricted. The requirements for escape routes, however, are stricter for housing than for offices. Adding extra stairs may be necessary, though most new office buildings have sufficient escape routes.

7.3.9   Location

The only location characteristic that could be said to be a veto criterion for residential conversion is if noise levels are too high to the façade, if odours prevail and if there is a prevalence of fine particles in the air.

If the requirements for low noise level and clean air are not met, then residential conversion is not feasible. Other location characteristics are less critical, depending on the target group and the combination of characteristics. However, other housing projects nearby are a 'soft factor' that influences the conversion potential. Fourteen of the fifteen conversion projects that were studied are located in established housing locations or mixed-use locations. Only the Churchill Towers project had a different type of location, on the edge of an industrial/logistics area, near a housing area. Conversions of buildings in industrial areas were not considered attractive by housing associations or developers, who are the primary actors initiating most conversions. However, residential conversions for specific target groups are possible in mono-functional office locations, that is, if the location is situated near the central business district and near areas with social and commercial facilities.

7.3.10 Building

Summarising the characteristics that have an effect on a building's conversion potential, there are few building characteristics that make conversion into housing impossible. A building is more easily adapted than its location. The characteristics of the structure and the floors are the most crucial for the conversion potential. The scale of the structure must allow separation into usable spaces. While older office buildings were not built according to standard measurements, office buildings from the 1980s onwards often have a structure that is a multiple of 1.8 m, such as 7.2 m, and is well suited for accommodating housing. A specific risk with older office buildings is that measurements and materials used do not always correspond to the construction drawings, and the plan sometimes differs between floors. Another potential risk with older buildings is poor maintenance and deterioration of the structure (e.g. corrosion of steel reinforcement in concrete).

The floors of office buildings normally provide enough strength for residential conversion. Problems may occur though when manipulating the floors. A typical floor in an office building is made of prestressed hollow core slabs. If the reinforcement in the floors is cut, the floors lose strength. Apartment buildings require a higher density of vertical shafts than office buildings. Penetrating the floors to create shafts for water, electricity and sewer is one of the problems of converting offices into housing. Though several building characteristics represent potential risks for the legal, functional, technical and cultural feasibility and thus also for the financial feasibility of conversion projects, only one characteristic represents a veto criterion: the floor-to-ceiling height must equal or exceed 2.6 m.

The characteristics of the façade influence the conversion potential of office buildings significantly. Though the façade is often adaptable, all adaptations imply extra building costs and hence reduce the financial feasibility of a conversion. As the requirements for thermal and acoustic insulation are higher for housing than for offices, adaptations of the façade are needed in most conversion projects.

Finally, the image of outdated office buildings does not always trigger positive reactions from potential residents. Though some office buildings are listed monuments or renowned buildings that have a specific image or are even able to provide a specific identity to a whole neighbourhood, most office buildings are very ordinary and have an image too strongly related to office work. In these cases, the facade is often replaced, even if it is technically well maintained and meets the requirements for housing. The location and building characteristics that influence the residential conversion potential of office buildings are summarised and presented as a checklist (see Table 7.4).

Table 7.4 Location and building characteristics that negatively influence conversion potential of offices into housing.

Location	Criterion
Functional	
Urban location	Mono-functional industrial or office park
Distance and quality of facilities	Shops for daily necessities >500 m
	Public meeting space (square, park) >500 m
	Restaurant/cafe >500 m
	Bank/post office >1 km
	Education, sports, basic medical facilities >1 km
Public transport	Distance to railway station >1 km
	Distance to bus/metro/tram >250 m
Accessibility by car and parking	Distance to parking place >100 m
	<1 parking place per housing unit
Cultural	
Status of neighbourhood	Situated near city edge (e.g. near motorway)
	No housing in immediate vicinity
	Poor or no public space in neighbourhood
	Area has poor reputation/image; vandalism
	Noise or stench (factories, trains, cars)
Legal	
Environment	Noise load on facade >50 dB (limit for offices 60dB)
	Level of fine dust above norm
Building	**Criterion**
Functional	
Features of new housing	<20 units of minimal 100 m² can be realised
	Unsuitable layout for selected target group
Extendibility	Not horizontally or vertically extendable
Technical	
Maintenance	Poorly maintained, deteriorating structure
Structure dimension	Building depth <10 m
	Structural grid <3.6 m
	Distance between floors <2.6 or >3.6 and <5.5 m
Support structure	In poor condition, not sufficient for housing
Facade	Maladaptive, impossible to attach interior walls
Installations	Impossible to fit vertical ventilation shafts
Cultural	
Character	Lack of identity or negative image
	Image not adaptable
Legal	
Environment	Poor acoustic insulation exterior and interior
	Poor thermal insulation
	Too little daylight; less than 10% of equivalent floor area
	No elevators in buildings higher than four floors
	No emergency stairways or not sufficient stairways

Adapted from Geraedts and Van der Voordt (2007)

7.4   Conclusion

Conversion to housing is a means of coping with obsolete buildings and takes place especially in city centres or in centrally located housing areas. From the case studies, the most important aspects that influence residential conversion potential of buildings were the:

- Segregation of real estate markets and actor roles
- Purchase price of buildings for conversion
- Location characteristics
- Building characteristics
- Conversion building costs

In the Netherlands, housing demand is structurally high and is predicted to remain high at least until 2025. In some locations housing rents are higher than office rents. In these locations, like in the centre of Amsterdam, residential conversion is especially attractive. Governmental subsidies are not a conversion trigger for any of the parties; rather, municipal cooperation on policies and legislation is found to be important. The price for obsolete office buildings is often too high for conversion to be feasible, while owners are not eager to sell buildings with financial loss. However, as the market value of an office building is related to its potential yield, investors need to consider realistic and convincing future yields to calculate a credible market value for structurally vacant properties. When refusing to devaluate, investors seem to have forgotten a basic principle from general economic theory: never consider the investments made, only their possible future yields.

In the Netherlands, 70% of the office buildings are located in monofunctional office locations. In general, the accessibility of these locations is good, both by car and by public transport. Possible scenarios for residential conversions in these areas are developments starting at the edges of the locations, developments in phases starting with the addition of facilities, the 'ink stain' development method or integral urban area regenerations. While the last option seems to be the most successful, it is also the most complicated. Given that complete office locations are reducing, such integral adaptations may prove more attractive over time.

The measurements and technical state of the building structure are critical building characteristics for conversion. Subsequently, building characteristics are considered that have a more direct effect on conversion building costs; these are characteristics of the façade and installations and costs related to the level of finishes of the new housing programme. As the building cost equals approximately 50% of the total initial investment cost, a good estimate of building cost is important. The most striking risks of conversions are legal or technical aspects, though these eventually translate into financial aspects. If taken into account, most risks can be dealt with, increasing the feasibility of a project. The 15 cases that were studied in this research show that it is possible to generalise the opportunities and risks of conversion, that is, the critical success and failure factors. Assessing these factors in the initial phase of a conversion project can contribute to the increased feasibility of conversions.

References

Flyvbjerg, B. (2006) Five misunderstandings about case-study research, *Qualitative Inquiry*, 12(2), 219–245.

Geraedts, R. and De Vrij, N. (2004) Transformation meter revisited, in *Open Building Implementation*, Kendall, S. ed., Munice, IN: Ball State University.

Geraedts, R.P. and Van der Voordt, D.J.M. (2007) A tool to measure opportunities and risks of converting empty offices into dwellings, in *Sustainable Urban Areas*, Rotterdam: ENHR.

Mason, J. (1996) *Qualitative researching*, London: Sage.

Miles, M.B. and Huberman, A.M. (1994) *Qualitative data analysis an expanded sourcebook*, Thousand Oaks, CA: Sage.

Patton, M.Q. (2002) *Qualitative research and evaluation methods*, Thousand Oaks, CA: Sage.

Remøy, H. (2007) De markt voor transformatie van kantoren tot woningen, in *Transformatie van kantoorgebouwen thema's, actoren, instrumenten en projecten*, Van der Voordt, T., Geraedts, R., Remøy, H. and Oudijk, C. eds., Rotterdam: Uitgeverij 010.

Remøy, H. and Van der Voordt, T.J.M. (2007) Conversion of office buildings; a cross-case analysis, in *Building Stock Activation 2007*, Kitsutaka, Y. ed., Tokyo: Tokyo Metropolitan University.

Yin, R.K. (1989) *Case study research: design and methods*, Thousand Oaks, CA: Sage.

Preserving Cultural and Heritage Value

8.1 Introduction

This chapter considers the value of cultural heritage. The theoretical framework of cultural heritage value has mainly been developed from the point of view of governmental bodies. For that reason, but also because of the difficulty of quantifying the cultural value in a monetary way, literature stressed the noneconomic benefits of cultural heritage. A new way of thinking is taking place in governmental policies, seeing heritage as an important mechanism to fulfil contemporary society's demand for cultural experience and leisure. This new way of thinking can be interesting with respect to privately owned heritage as well, focusing on the return on investment. Heritage value is a complex concept comprising variously defined values. Three main categories will be described here: use values, aesthetic values and indirect values.

8.2 Historic Heritage

Historic heritage has always been considered 'irreplaceable and precious' (ICOMOS 1999), enriching people's lives, often providing a deep and inspirational sense of connection to community and landscape, to the past and to lived experiences. Heritage buildings are historical records particularly important as tangible expressions of cultural identity (Productivity Commission 2006). Places of cultural significance reflect the diversity of our communities, telling us who we are and how the past has helped form our lives and our surroundings. These places are irreplaceable and precious. For all those reasons national governments and international bodies in the past

Sustainable Building Adaptation: Innovations in Decision-Making, First Edition.
Sara J. Wilkinson, Hilde Remøy and Craig Langston.
© 2014 John Wiley & Sons, Ltd. Published 2014 by John Wiley & Sons, Ltd.

tried to develop strategies, methodologies and criteria to define and preserve historic heritage, on behalf of the cultural value that they represent. In that sense, the definition of cultural and natural heritage provided by UNESCO in the World Heritage Convention appears particularly interesting, considering properties to have cultural and natural relevance when having outstanding universal value (World Heritage Centre 2008). Outstanding universal value means cultural and/or natural significance that is as exceptional as to transcend national boundaries and to be of common importance for present and future generations of all humanity. As such, the permanent protection of this heritage is of the highest importance to the international community as a whole.

The committee defines the criteria for the inscription of properties on the World Heritage List. Furthermore, in Articles 1 and 2 of the convention, UNESCO presents a list of properties, which may be considered cultural and natural heritage. Those are directly derived by the experiences of the Athens Charter (ICOMOS 1931), the Hague Convention (ICOMOS 1954), the Venice Charter (ICOMOS 1964) and the Nara Document on Authenticity (ICOMOS 1994). UNESCO also mentions the concept of mixed cultural and natural heritage (World Heritage Centre 2008), defining a particular category of properties satisfying a part or the whole of the definitions of both cultural and natural heritage laid out in Articles 1 and 2 of the convention. Strictly connected with this the idea of cultural landscapes has been introduced, referring to cultural properties which represent the combined works of nature and man designated in Article 1 of the convention. The cultural landscapes are illustrative of the evolution of human society and settlement over time, under the influence of the physical constraints and/or opportunities presented by their natural environment and of successive social, economic and cultural forces, both external and internal. Historic heritage places may generate benefits in the way they are utilised. Beyond this direct value, there is also the potential for historic heritage places to generate indirect value by cultural benefits (Productivity Commission 2006).

8.3   The Value of Heritage

8.3.1   The Value of Place

The Burra Charter links the heritage value of a place to the cultural significance of a site (ICOMOS 1999). This concept has the disadvantage of being extremely subjective and different across countries, depending on community values and expectations (Productivity Commission 2006). However, the cultural importance of heritage has been stated by many researchers. Heritage has been defined as expression or representation of the cultural identity of a society in a particular period (Koboldt 1995), as well as contribution to the community's cultural capital (Throsby 1997). Norberg-Schulz (1980) related the identity, the genius loci, of a place directly to its history and meaning, whereas Augé (2000) speaks of historical events and usage as what creates a symbolised place. In that sense the cultural value can be defined specifically in the context of built heritage, as the value that can be attributed to a building, a collection of buildings or a monument. This cultural value is additional to the value of the land and buildings as purely physical entities or structures and embodies the community's valuation of the asset in terms of its social, historical or cultural dimension. Unfortunately, the identification of heritage deals inherently with subjective aspects; the same disadvantage can be found in classifying the degree of cultural significance. On the other hand, cultural values may be shared between different groups in society and at different scales at no added cost. For instance, a place considered to be culturally significant to a local community may also be regarded to have heritage value to a region, to a state or even nationally (Productivity Commission 2006). The heritage value on a national or regional scale is defined and guarded by national bodies like the Dutch Cultural Heritage Agency, English Heritage, Australian Heritage Council and US Heritage Foundation.

8.3.2   Cultural Capital

Until now the importance of heritage has been analysed from a merely cultural point of view, but the next step is to look at the question from an economic point of view as well. This is not an easy task, especially because the fundamental emotional and spiritual aspects belonging to the artistic experience or the power of the genius loci expressed by the sublime qualities of a place are generally perceived as something completely incompatible with economics. This conflict is experienced every day in practice, where arguments for heritage preservation are usually based on archaeological, architectonic or artistic (expert) value assessment, not an economic interpretation of value (Ruijgrok 2006). It is undeniable that the economic aspects have a significant dimension in the heritage decision process, but on the other hand, resources for the maintenance of cultural buildings and sites are not unlimited. Far harsher is the clash in situation where financial revenues, brought in by the use of old buildings, must be offset against possible damage to culturally significant property. Such trade-offs are familiar to economists. That is the reason why researchers started to question whether there are economic tools that can be useful for looking at heritage decisions: one of those is the concept of cultural capital (Throsby 2006). Economists look upon capital both as a store of value and as a long-lasting asset that produces a stream of services over time. An item of cultural heritage, such as a historical building, can be thought of as just such an asset. But its distinguishing characteristic as a specifically cultural capital good is that it embodies or yields not only economic value through its financial worth and through the economic services it provides but also cultural value through its historical or aesthetic significance and the cultural experiences it provides for the community.

For several reasons, it is becoming increasingly apparent that the concept of cultural capital can be helpful in analysing heritage and in formulating heritage policy. First, the definition of heritage as capital enables the related concepts of depreciation, investment, rate of return, etc., to be applied to the evaluation and management of heritage. In that way a profitable discussion can be opened between professionals whose job it is to care for cultural assets and economists who are concerned with the formulation of economic and cultural policy. Second, the idea of cultural capital is related to several specific forms of value, in particular to the cultural value as something distinct from economic value. Third, since capital assets are long lasting, the notion of cultural capital naturally leads to thinking about sustainability. That is quite important considering that environmentally or ecologically sustainable development is an area of economic growth. Those are aimed to preserve the natural resources of the planet for future generations and hence, speaking of culturally sustainable development, to safeguard our cultural heritage for the benefit of future generations. Last, defining heritage as cultural capital opens up the possibility of looking at heritage projects using a cost–benefit analysis. In that sense an intervention involving expenditure of public or private funds can be seen as a capital investment project. If the asset is a historical building or location, treating the cultural resource as an item or items of cultural capital enables the tools of financial investment appraisal to be applied, defining value as the amount of welfare that heritage generates for society. The definition of welfare is broad, and this valuation is significantly different from ordinary cost–benefit analysis also in the time stream of both economic and cultural value that is being evaluated and assessed. In other words, the identification of cultural value alongside the economic value generated by the project means that the economic evaluation can be augmented by a cultural appraisal carried out parallel, as an exercise comparing the discounted present value of the time streams of net benefits with the initial capital costs (Throsby 2006).

8.3.3   Benefits of Heritage Conservation

The conservation of heritage generates several benefits. Those can vary from commercial advantages (like tourism) to intangible community benefits (including the sense of history, educational and research value, spiritual value) (Productivity Commission 2006). Conservation can also be considered a duty to future generations. Cultural capital can be seen, like the physical capital in which it is contained, to be subject to decay if neglected. Existing cultural capital can have its asset value enhanced by investment in its maintenance or improvement; new cultural capital can be created by new investment. If these interpretations are accepted, the social decision problem in regard to this type of cultural capital might be seen within the framework of social cost–benefit analysis (SCBA) and approached by ranking projects according to their social rate of return (Productivity Commission 2006).

8.4   Assessing Economic Value of Heritage

8.4.1   The Market Value of Heritage

The economic value of cultural heritage is defined as the amount of welfare it generates for society, comprising both material and immaterial values (Ruijgrok 2006). A good example of this from Beijing is provided in Figure 8.1. Three different kinds of economic benefit can be defined: a housing comfort value, a recreation value and a bequest value. Recent development in environmental economic theory and social survey methodology has made it theoretically defensible and practically feasible to measure these values (De la Torre 2002). The methods used are such as hedonic pricing models (revealed-preference techniques) or contingent valuation methods (stated-preference technique), which are commonly used in cost–benefit evaluations and which also apply to measuring the economic value of heritage (Navrud and Ready 2002). In some cases, the illiquidity of heritage and the absence of a market are solved by a revealed-preference study of surrogate markets. Bear in mind that these methods may not comprise the cultural value fully, but they definitely show the willingness to pay for living in, visiting, saving and admiring heritage. Determining the economic costs and benefits of heritage creates opportunities for supporting investments in heritage for private and public parties. Moreover, it becomes possible to include the costs of heritage loss in cost–benefit evaluations for new developments in historical locations. Eventually, the market value of heritage can be defined.

Figure 8.1   Former machine factory, Beijing, China. Photograph provided courtesy of Dr. Hielkje Zijlstra.

Generally speaking, obsolete monuments have a low market value. The market value of a monument as well as newer buildings is determined on the basis of its value in use and the willingness to pay for this use. The market value of a building can be determined by assessing its future returns, referring to more than just the potential rental income: the market value can be divided into (a) direct market value, based on the value in use and sale or rent value, and (b) indirect market value, based on experience value and public relations (PR) value.

8.4.2   Direct Market Value

Both the utility (e.g. ticket) and sale or rent value can be expressed in monetary terms and are therefore regarded as the direct value of a monument. One way to determine this value is simply to subtract the investment costs from the income – determined by means of GIY. The value is determined on the basis of the new function.

The experience and PR values are not deducted directly from a project and hence represent the indirect value. Private parties tend to focus on direct values, while public parties also consider the indirect value. Studies have proven that heritage may influence the stream of tourists visiting a city, indicating the value of heritage for PR or city branding. The experience value of heritage is in the eye of the beholder: this value varies and depends on the experience and reference frame of the observer.

8.4.3   Indirect Value

The indirect value of a monument is based on its influence on spatial quality. The presence of an iconic monument affects its surroundings. A monument may have positive effects on tourism, economic structure, the labour market, preservation of specific competencies, living environment and the working environment. In what way and to which extent depend also on its use. Spatial quality is an important criterion for people and organisations deciding on new accommodation, and the social importance of spatial quality is recognised. The architectural quality of a monument, the quality function it is accommodating and the quality of its surroundings contribute to social benefits, like reduction of vandalism and increased (social) safety, again contributing to the market value of the building and its surroundings.

Hence, the characteristics of spatial quality can be connected to economic, social, cultural and environmental dimensions of society. Sooner or later obsolescent buildings and locations lead to extra costs in public sectors and influence municipalities' budgets.

Expressing the indirect value as financial value is challenging. It is obvious that heritage has a positive influence on the value of a building and its surrounding, but it is not so easy to deliver 'hard evidence'. SCBA can be used to reveal the costs and benefits of all related parties and actors. To express the SCBA in money, different economic valuation methods were developed to decide this value. Next to a market analysis, six other methods are used to express the economic value of spatial quality (Dammers et al. 2005):

- *Prevention-cost method*: used to define costs and benefits from, for example, vandalism prevention
- *Averting behaviour method*: used to calculate the costs that households employ to prevent a decline in the quality of their direct living environment
- *Recovery costs*: used to calculate the costs of interventions that serve to prevent or compensate a decline or loss of quality as a result of a low-quality development in the surroundings
- *Hedonic pricing method*: defines the value of spatial quality by willingness to pay, measured by the actual selling or rental prices of real estate
- *Travel cost method*: uses the travel costs that people are willing to make to visit a specific location or building as a measurement for the economic value of the location or building
- *Contingent valuation method*: defines the value of spatial quality by the willingness to pay, measured by a survey

8.4.4   Indirect Value of Heritage Tourism

Heritage has the capacity to create strong indirect benefits. One of these indirect benefits concerns the positive consequences of cultural tourism, which is able to generate a twofold expenditure multiplier process. The first is related to the income of people working in the cultural industry, who will benefit from activities attracting tourists. The second refers to the resources spent by tourists for accommodation, food and leisure activities, secondary to the main tourist attraction. However, very often these indirect benefits are overestimated, and the differences in multipliers used are substantial and not easily validated. This is related to the fact that the effects of 'leakage' are neglected: one thing is to recognise expenditure and income potential, and another thing is to consider if the benefits of such expenditures would be local; many of these expenditures can be a benefit for nearby areas, something which considerably reduces the local benefit. An important conclusion therefore appears: in order to increase the heritage leverage, it is necessary not only to attract tourists and expenditure flows but to make tourists spend money on local products (Lusardi 2011).

8.4.5   Heritage as a Source of Skills and Competencies

Heritage may be considered as a source of developing skills and competencies. This is an indirect use value particularly interesting because of the possible impact on the local economy. An example of that is the workshop school in Spain, the 'escuelas taller' (Greffe 1998). When a decision is taken in order to rehabilitate or conserve a public monument, a historical site or even a garden, the promoters organise a school that will exist only until the completion of the public works. Usually young people without work or specific qualifications are recruited as both employees and students. They will benefit during this contract from practical and theoretical training, and at the end of this contract, they are expected to make other economic sectors and activities benefit from the skills and competencies they have developed. This workshop-school system may therefore satisfy three objectives: the rehabilitation of new heritage, the reproduction and dissemination of traditional skills and competencies and an increase of the quality of future works in other sectors of the economy, mainly in the fields of housing and urban development.

8.4.6   Private/Public Value

One very important distinction that has to be highlighted in order to better understand the value of cultural heritage is that between private and public value, or in other words between private and public interest. Private actors do not necessarily derive cash flows from conserving heritage, since the benefits like recreational perception values and bequest values accrue to the general public (Ruijgrok 2006). In the heritage arena this distinction is seen most obviously in the listing process. Listing has direct impacts on private owners of heritage properties, for example, through costs of regulatory compliance or through development opportunities. At the same time listing affects public value; indeed, the essential intention of the listing process is to protect the indirect benefits of heritage as experienced by the public at large. The main question is: who benefits and who pays? In that perspective a cost–benefit analysis is the most appropriate methodology to evaluate the costs and benefits with special regard to noneconomic factors. The private/public value distinction is recognised in the wider field of investment appraisal where the differences between private and social CBAs are well understood. In the case of a heritage project such as the conservation or adaptive reuse of a privately owned property, a CBA undertaken from a private viewpoint would use the actual financial flows and opportunity costs as experienced by the individual owner (Lusardi 2011). An SCBA of the same project would adjust the private analysis to account for taxes and transfers; use shadow prices, not market prices; use a lower discount rate reflecting a social time preference rate; include all indirect market effects (public goods and externalities); and recognise, if possible, any cultural value or collective benefit not otherwise accounted for. Once such adjustments are considered, private and social rates of return can be compared as a basis for the decision-making process.

8.5   Heritage Value and Adaptation

When a building reaches the end of its functional or economic lifespan, it will be assessed and a strategy for adaptation for new use or demolishment will be drawn up. Adaptation of a building takes place when one or more players are aware of the (potential) qualities of a building and/or its environment. Building conversions are only viable when the involved actors have a vision about the future potential of the building in its urban context, in the light of urban and social developments planned. In the considerations whether a building will be reused and transformed or not, the cultural and heritage value of the building plays an important role. The architectonic, heritage and historical values of a building are weighed against the potential use value and the financial value (Roos 2007). The potential use and financial values are important for newer office buildings and the typical '13 in a dozen' buildings, while cultural–historical values become important and sometimes play the critical role when redeveloping historical and heritage buildings. The cultural and heritage value of a building is manifest in the building itself but also in combination with the significance of the building's history. The cultural and heritage value can be subdivided and described as aesthetic, emotional, experience, architectonic and cultural–historical value. The aesthetic value is seen as part of the architectural value. Throughout history, the aesthetic value of buildings has been discussed extensively, mainly by architectural writers (Le Corbusier 1986; Venturi et al. 1996; Vitruvius and Rowland 2003), and lately also the value of aesthetics or architectural quality has been studied (Fuerst et al. 2011).

8.6   Architectonic and Aesthetic Value

Heritage has the ability to trigger strong feelings. Heritage is used in the branding of cities and plays with the interest people have towards discovery and knowledge of the artistic characteristics and historical progressions that characterise a specific building period. These values and the corresponding services are more and more linked with the explosion of cultural tourism (Greffe 1998).

The seed silos in Islands Brygge in the harbour of Copenhagen were radically adapted into high-quality, high-price apartments (see Figure 8.2), following the design of the world-famous architects MVRDV. The apartments were added on the outside of the original silo walls, resulting in apartments with free floors, glass facades and great views across the harbour. The spectacular building shape expresses the historic development of the building and the area. In Berlin, a former custom and tax authorities building was adapted and extended (see Figure 8.3 and Figure 8.4). The architectural design of Sauerbruch Hutton architects adds a spectacular new appearance.

Heritage may play an active role in educating young and older generation. This is connected with the original functions of museums that were specifically created to train artists. But this function can also be played by other types of heritage, like monuments, collections and industrial heritage, which can express the same value as museums. In this context, it is particularly interesting that the progression of the multimedia has allowed delivering of new tools and products that increase the quantity and quality of this potential learning through heritage. Heritage is a good illustration of what is now recognised as 'edutainment' (Greffe 1998). An example from Turin is provided in Figure 8.5.

Figure 8.2   Seed silos in Islands Brygge, Copenhagen, Denmark. Photograph provided courtesy of Dr. Juriaan van Meel.

Figure 8.3   Feuerwache Regierungsviertel (exterior), Berlin, Germany. Photograph provided courtesy of Oliver Schäffler.

Figure 8.4 Feuerwache Regierungsviertel (detail), Berlin, Germany. Photograph provided courtesy of Oliver Schäffler.

Figure 8.5 Media Centre for Olympic Games, Turin, Italy. Photograph provided courtesy of Dr. Hielkje Zijlstra.

8.7   Experience Value

The experience value is comprised in the emotional value, and the two largely depend on each other. Experience value is the value that an individual or a group in a certain time and in a given context attributes to a building. The experience value may be decisive for preservation of a building, especially when the building is found functionally fit, though financial profit by conversion is uncertain and conversion only takes place after lobbying and campaigning by individuals or groups. The meaning of the term experience value and how to measure it is one of the questions frequently asked when deciding about building adaptation. Not only the building itself is important to answer this question but also the relationship between the building and its surroundings (how the building reacts on its surroundings, which effects the surroundings have on the building).

Experience value has many aspects, and everyone looks at it differently depending on their background. An unambiguous definition of experience value is not possible because of the nuances it comprises and the specific situation of each building individually. However, specifying the several aspects of experience value makes the concept more meaningful. Benraad and Remøy (2007) distinguished seven different aspects: familiar ugliness, cultural historic value, symbolic value, traumatic experience value, value in use, intrinsic value and the relationship between building and location.

8.7.1   Familiar Ugliness

Many buildings that have been transformed were retained because the buildings were considered important, without being aesthetically important. In Eindhoven 'the White Lady' (former Philips light bulb factory) was initially nominated for demolition. Most people found the building ugly. However, artists pleaded for reuse of part of the former Philips buildings, including the White Lady. Ultimately, the building was refurbished, and it even determined the development of the area around it (see Figure 8.6). The intrinsic value of the building was not automatically recognised and acknowledged. The apparent ugliness of the building proved to be important in this case and could be significant for other buildings in the urban context.

Original year of construction: 1926
Year of conversion: 1998
Floor space after conversion: 36,460 m²
Developer: De Witte Dame v.o.f.; a combination of the two developers
    Van Straten Bouw and IBC Bouwgroep
Architect: Diederendirrix, Eindhoven, http://www.diederendirrix.nl
Constructor: Van Straten Bouw bv and IBC Bouwgroep

8.7.2   Cultural–Historical Value

The cultural–historical value of a building is another reason to keep a building. A building is part of the history of the city. This does not mean a priori that every old building with specific style is of historical interest.

Figure 8.6   The White Lady, Eindhoven, the Netherlands.

The uniqueness plays along. The Royal Palace at 'the Dam' in Amsterdam has a cultural and historical value (see Figure 8.7). It was built by patricians who in their time were the world's richest people. They contracted the best architects of their time. This building has both a historical and an architectural value. Another example of cultural–historical value is the value of representation of historical building types. For instance, value is ascribed to the last remaining farm building from a specific period by the uniqueness of the particular typology and building tradition. Yet, change through demolition and new construction is an essential part of a dynamic spatial development. For reuse, the issue is the additional or specific qualities of the building and its context compared to new construction.

8.7.3   Symbolic Value

Symbolic value can be distinguished from the cultural and historical value. The large church in Veere, a village in the southwest of the Netherlands, is now a listed monument. The church was completed in 1521. The building was too large for the community who used it but was constructed this big so that everyone in the village would fit. The church lost its function by the church reformation in 1537. In 1686 it was ravaged by fire. In 1811 the church was taken into use by Napoleon as a military hospital, after which the building no longer had a function. In order to prevent destruction, the Great Church of Veere was purchased by the state in 1881 as the first Dutch national monument. Since then the monument is the responsibility of the Government Buildings Agency.

Figure 8.7   The Royal Palace, Amsterdam, the Netherlands.

The church is a symbol of the history of glory and decay. It is part of history, painters have painted it, and poets have written about it. The building maintenance is costly. In the 1990s conversion into housing was considered and eventually rejected. The building is a symbol of the church and its ambition in the fourteenth century, and a private function was not found suitable. A conversion of the church was commissioned by the Government Buildings Agency, and it was completed in 2004 as a stage for the Foundation for New Music, designed by Marx and Steketee Architects (see Figure 8.8).

The design brief was on the one hand to keep the almost untouchable monument intact and let the eventful history of the Great Church 'speak' and on the other hand to accommodate a new cultural use with constantly changing configurations. The entrance was kept at its original place under the tower with a newly added entrance portal. The north and south aisles on either side of the entrance feature spaces for ticketing, coffee bar and souvenir store. The new architectural intervention in the transverse nave consists of a flat concrete floor field with two steel columns on either side of the stage. This construction bears timber floors that are a reconstruction of a part of the French-built hospital floors that were demolished around 1880. These floors now carry service units for the podium function of the Foundation for New Music: changing rooms, practice rooms, offices, restrooms, storage and technical rooms. The architectural composition of girders, beams and objects is not only a utilitarian solution to a functional problem but also a tool for making the tormented history of the Great Church visible.

Figure 8.8   The Great Church, Veere, the Netherlands.

8.7.4   Traumatic Experience Value

A building may be negatively valued by activities that happened at that location. These experiences do not necessarily have anything to do with the building itself, but are often projected on the building. This shows how subjective experience value is. An example is the Maupoleum in Amsterdam (see Figure 8.9). The building had a prominent place on the list of ugliest buildings in Amsterdam. This was partly because it is one of the most criticised buildings in town, which had nothing to do with the building itself, but the way it was operated. The owner – Maup – demanded so much rent from the Jewish textile merchants who settled there that they went bankrupt. Due to such a poor treatment of Jewish tenants after the horrors of World War II, the building attracted negative reputation. Twenty years after completion it was demolished in 1994. The monument commission tried to save the building for conservation, because with its strong horizontal lines and concrete facade, it had a cultural–historical value as an exponent of the architecture of the 1960s of the twentieth century. However, the negative traumatic experience value was so strong that only demolition offered a solution. In some cases however, the traumatic experience value of a building or a location is so high that demolition is not acceptable and a state memorial or museum or a combination of the two is the only possible development. Examples of such buildings and locations are the Tower of London and, more recently, the Ground Zero site (former World Trade Center) in Manhattan, New York.

Figure 8.9   The Maupoleum, Amsterdam, the Netherlands.

8.7.5   Value in Use

The utility affects how a building is experienced. If it no longer meets the requirements for use of this time, it is functionally obsolete and the experience value becomes dependent on the use value. An example is the former Permeke Ford Car showroom in the centre of Antwerp (see Figure 8.10 and Figure 8.11), dating back to 1925. After more than 50 years on the same location, the success of the showroom made it necessary to move – because of the density and congestion of the city centre, it was impossible to expand the showroom. The building was neglected and left vacant for 15 years before the municipality bought it and adapted it into a mixed-use building comprising a public library, municipal offices and a restaurant. After 15 years of vacancy and decay, the experience value of the building and location was negative. The large, diverse and separated spaces with daylight admission from above and to the north eventually made this change of use adaptation possible. The building is now perceived as a building with high architectural value.

8.7.6   Intrinsic Value (Highest and Best Use)

The intrinsic value of a property is the value of the building itself. When adapted, the intrinsic value is the possibility for new use of the building. A building will be converted when it has the potential to absorb new functions. The possibility of functional change may also be referred to as the capacity to change or the capacity of use of a building. The intrinsic value or worth of a building is practical. The intrinsic value may also relate to parts of the

Figure 8.10 The Permeke Library (exterior), Antwerp, Belgium. Photograph provided courtesy of Dr. Hielkje Zijlstra.

Figure 8.11 The Permeke Library (interior), Antwerp, Belgium. Photograph provided courtesy of Dr. Hielkje Zijlstra.

building. If the structure of a building is useful but not the façades, then the structural part could be reused. The structure then has intrinsic value. The vision of the (future) uses of the building in many cases determines whether resources can be deployed to build forth on the intrinsic value. Intrinsic value is provided for experience value, but cannot be seen apart from the experience value. The experience aspects in Section 8.7 are highly subjective and depend on the person or group and the zeitgeist, the subjectivity of the moment. Quite often, the intrinsic value of the building only becomes visible after conversion and quite often depends on the vision of an individual or a group. The intrinsic value is in this sense equal to the highest and best use.

8.7.7   Heritage as a Source of Social Value

Heritage communicates collective and social values and references for new construction and managing of existing properties. Its influence is disseminated throughout the society. The effect of heritage on its surroundings depends on its use. Activities related to heritage may be used to support a learning and integration process for new generations, by the creation of museums or simply by the revival of cultural craftsmanship related to restoration and adaptation of heritage buildings. In many European countries several practices have been tried in that sense. Heritage can be used to redefine locations. Many recent urban redevelopment strategies were based on cultural components like the rehabilitation of an old historical city centre or redevelopment of an urban area. Over time the message is more or less the same: this cultural revival is a way of adding value to the environment; the cultural revival shows that it is possible to recreate in an environment where creation had already been organised (Lusardi 2011). Examples from Florence and Dublin are provided in Figure 8.12 and Figure 8.13 respectively.

Adaptation of a building may have an effect for a variety of developments in the area. The redevelopment of a factory building in an impoverished location can reverse the downward spiralling development of that area. It can convince potential investors to convert or adapt other buildings or to proceed to demolition and new construction, with or without change of function. Converted buildings carry on the identity of the neighbourhood. In the vision of area development, conversion can also mean temporary new use of a building, pending new plans. The building, and in particular its new function, serves as a pioneer for a desired or planned development. A good example is the North Atlantic House in Copenhagen, a former warehouse that was used for fish storage for over two centuries (see Figure 8.14 and Figure 8.15). In the beginning of the 1990s, it was adapted to house several research and cultural institutes concerning North Atlantic culture and history and a famous restaurant for Nordic food.

Once this development is put in motion and the desired effect is achieved, the building can be more permanently converted or may be demolished to make place for new construction. There are also situations where conversion does not contribute to the experience value of an area. The relationship between building and area is always a symbiosis. If only one historical

Figure 8.12 Former prison Murate, Florence, Italy. Photograph provided courtesy of Dr. Hielkje Zijlstra.

Figure 8.13 Former harbour warehouses adapted to shopping mall, Dublin, Ireland. Photograph provided courtesy of Dr. Hielkje Zijlstra.

Figure 8.14   The North Atlantic House (exterior), Copenhagen, Denmark. Photograph provided courtesy of Dr. Juriaan van Meel.

Figure 8.15   The North Atlantic House (detail), Copenhagen, Denmark. Photograph provided courtesy of Dr. Juriaan van Meel.

building remains in the existing urban situation, it may be better to demolish this building also. Otherwise, the building is alienated from its location and adds nothing to the new situation. The relationship between building and location is always present. Especially in conversion or redevelopment processes of areas, this relationship should be properly safeguarded. The existing buildings' experience value adds to the spatial context. The functions that are assigned to existing buildings are of importance. Conversion of existing buildings can be used as a trigger for redevelopment. This allows the specific identity to become carrier of the general.

An example of a successful transformation of an inner-city brownfield location is the Vulkan area in Oslo. The area is located near the river Akerselva, which was the location for several industrial premises. The location accommodated industrial functions, that is, a wood mill, long before the industrial revolution. In 1873, an iron foundry was located on the location that was in use until the end of the 1950s. From the 1960s onwards, the buildings were used for several different industrial purposes and as office space. Because of the topography of the location, the size of the buildings and the historic industrial use, the location formed a sealed-off area in the city, unknown to the public. In 1996 Statsbygg, the Norwegian Government Buildings Agency, bought the location for use as theatre production and storage location. This project was never realised. The location and buildings were thereafter sold several times, before Vulkan Property bought the location. Oslo-based LPO architects designed the plan for a mixed-use redevelopment, preserving and adapting some of the buildings in the location, which was approved by the municipality in 2009, and now the location is transformed into a new city centre neighbourhood. Central in the redevelopment are historical buildings that now accommodate an event hall, a dance scene and a food hall, Mathallen. Mathallen (see Figure 8.16 and

Figure 8.16   Mathallen (exterior), Oslo, Norway. Photograph provided courtesy of LPO architects.

Figure 8.17 Mathallen (interior), Oslo, Norway. Photograph provided courtesy of LPO architects.

Figure 8.17) is designed following examples of southern European food halls, accommodating different bars, cafés and restaurants and offering several shops for local food and food specialties. The opening of the hall made the area accessible to the public and completed the transformation of this centrally located area. The reuse of the industrial building connects the new use to the history of the area, giving it a specific identity different from the surrounding areas. It is an example of heritage adaptation where public and private interests are fully represented, conserving cultural value and enhancing economic value.

8.8   Conclusion

The theoretical framework of cultural heritage adaptation has been developed mainly from the point of view of governmental bodies. For that reason, but also because of the difficulty of monetary quantification of the cultural value of a building, literature has stressed the noneconomic benefits of heritage. A cost–benefit analysis may be a good tool for implementing noneconomic parameters with financial considerations. Heritage value can be measured in monetary terms based on the housing comfort value, the recreation value and the bequest value. The methods used to measure these include hedonic pricing models or contingent valuation methods. On the other hand, a new way of thinking is taking place in governmental policies and commercial heritage redevelopment, tending to see heritage as important asset to fulfil the demand for cultural experience and leisure coming from contemporary society. In that sense heritage can express an important role, creating spillover benefits for tourism and other activities. This new way of thinking can be interesting with respect to privately owned heritage focusing on return on investment. Cultural value expressed by heritage can lead to higher rental rates and occupancy levels. Cultural value, and specifically experience value, is also an important factor in the assessment of adaptive reuse potential. A building with high experience value has the capacity to define the whole location surrounding it, making adaptive reuse of not only the building but also conversion of the location financially attractive. This chapter has shown that cultural value is not covered by a univocal definition, but is rather a complex concept, comprising several partial definitions.

References

Augé, M. (2000) *Non-places introduction to an anthropology of supermodernity*, London: Verso.

Dammers, E., Hornis, W. and Heemskerk, I. (2005) *Schoonheid is geld!: naar een volwaardige rol van belevingswaarden in maatschappelijke kosten-batenanalyses*, Rotterdam: NAi Uitgevers.

De la Torre, M. (2002) *Assessing the values of cultural heritage*, Los Angeles, CA: Getty Conservation Institute.

Fuerst, F., McAllister, P. and Murray, C.B. (2011) Designer buildings: estimating the economic value of signature architecture, *Environment and Planning A*, 43, 166–184.

Greffe, X. (1998), *The economic value of heritage*, Colloquio di Ahmedabad, National Trust of India.

ICOMOS. (1931) The Athens Charter for the restoration of historic monuments – 1931. Available at http://www.icomos.org. Accessed on 8 August 2013.

ICOMOS. (1954) Convention for the protection of cultural property in the event of armed conflict. Available at http://www.icomos.org. Accessed on 8 August 2013.

ICOMOS. (1964) International Charter for the conservation and restoration of monuments and sites (The Venice Charter). Available at http://www.icomos.org. Accessed on 8 August 2013.

ICOMOS. (1994) The Nara Document on authenticity. Available at http://www.icomos.org. Accessed on 8 August 2013.

ICOMOS. (1999) *The Burra Charter: the Australia ICOMOS charter for places of cultural significance*, Melbourne: T.A.I. Incorporated. Melbourne.

Koboldt, C. (1995) *Optimising the use of cultural heritage*, CSLE Discussion Paper No. 96-01, Saarbrücken: Saarland University.

Le Corbusier (1986) *Towards a new architecture*, New York: Dover.

Lusardi, I. (2011) Torre Velasca: determining the highest and best use of a modern monument, Master's Thesis, Delft University of Technology, Delft, The Netherlands.

Navrud, S. and Ready, R.C. (2002) *Valuing cultural heritage: applying environmental valuation techniques to historic buildings, monuments and artifacts*, Cheltenham: Edward Elgar.

Norberg-Schulz, C. (1980) *Genius loci: towards a phenomenology of architecture*, London: Academy.

Productivity Commission (2006) *Conservation of Australia's historic heritage places*, Canberra: The Productivity Commission.

Remøy, H. (2007) De markt voor transformatie van kantoren tot woningen, in *Transformatie van kantoorgebouwen thema's, actoren, instrumenten en projecten*, Van der Voordt, T., Geraedts, R., Remøy, H. and Oudijk, C. eds., Rotterdam: Uitgeverij 010.

Roos, J. (2007) *Discovering the assignment/de ontdekking van de opgave*, Delft: VSSD.

Ruijgrok, E. (2006) The three economic values of cultural heritage: a case study in the Netherlands, *Journal of Cultural Heritage*, 7(3), 206–213.

Throsby, D. (1997) Seven questions in the economics of cultural heritage, in *Economic perspectives on cultural heritage*, Hutter, M. and Rizzo, I. eds., New York: St. Martins Press.

Throsby, D. (2006) Paying for the past: economics, cultural heritage, and public policy. Joseph Fisher lecture delivered on 16th August, 2006, School of Economics, The University of Adelaide..

Venturi, R., Izenour, S. and Scott-Brown, D. (1996) *Learning from Las Vegas: the forgotten symbolism of architectural form*, Cambridge, MA: MIT Press.

Vitruvius, P. and Rowland, I.D. (2003) *Ten books on architecture: the Corsini incunabulum with the annotations and autograph drawings of Giovanni Battista da Sangallo*, Roma: Edizioni dell'Elefante.

World Heritage Centre (2008) *Operational guidelines for the implementation of the World Heritage Convention*, Paris: UNESCO.

Part III Adaptation Decision-Making and Optimisation

The author for this part is Dr Craig Langston. Craig is Professor of Construction and Facilities Management at Bond University's School of Sustainable Development, Gold Coast, Australia. He has a combination of industry and academic experience spanning more than 35 years.

The research described in this part is the result of three Australian Research Council (ARC) Linkage Project grants comprising:

- *2009–2012 – Langston, C., Smith, J., Herath, G., Datta, S., Doloi, H. and Crawford, R.H., Making Better Decisions about Built Assets: Learning by Doing, ARC Linkage Project $180,000 LP0990261 (Industry partners: Williams Boag Architects and Assetic Australia).*
- *2007–2010 – Langston, C., Liu, C., Beynon, D. and de Jong, U., Strategic Assessment of Building Design Adaptive Reuse Opportunities, ARC Linkage Project $210,000 LP0776579 (Industry partners: Williams Boag Architects and The Uniting Church in Australia).*
- *2006–2009 – Crawford, R.H., Datta, S. and Langston, C., Modelling Environmental and Financial Performance of Construction. Sustainability Innovation Feasibility Tool. ARC Linkage Project $179,000 LP0667653 (Industry partner: Williams Boag Architects).*

Craig is author of five international books. In 2010, he also won the Bond University Vice-Chancellor's Quality Award for Research Excellence. He was awarded the Emerald Literati Network Award for Excellence in 2013 for his paper 'Validation of the adaptive reuse potential (ARP) model using iconCUR', Facilities, 30(3–4), 105–123 (2012).

This part deals with the problem of making effective decisions concerning building adaptation, not just from an economic perspective, but being

Sustainable Building Adaptation: Innovations in Decision-Making, First Edition.
Sara J. Wilkinson, Hilde Remøy and Craig Langston.
© 2014 John Wiley & Sons, Ltd. Published 2014 by John Wiley & Sons, Ltd.

cognizant of social and environmental objectives that collectively define the characteristics of a worthwhile project. Adaptation is an intervention to an existing physical state, and raises issues of appropriate timing, prioritisation and potential value add. In the latter case, this may lead to proposals that comprise a change in functional purpose, heritage preservation strategies, renewal or revitalisation of components, buildings or precincts, and/or significant improvements in operating performance such as lower energy demand, water usage and other environmental impacts.

Making effective decisions entails considering multiple criteria. This commonly involves trade-offs not just between economic, social and environmental objectives but also within the detail of the proposed intervention and across a spectrum of project stakeholders. These decisions have short, medium and long-term implications that add another layer of complexity to the process. Some criteria are quantitative and can be measured objectively, while others are qualitative and sometimes intangible. The ability for any intervention decision to be optimised so that it reflects the best combination of the aforementioned is obviously a daunting prospect. In the end, perhaps what is sought is a balanced compromise that maximises the benefits while minimising the negatives to provide the highest ratio within the available decision time frame.

So optimisation is a bit of a misnomer in this context. Nevertheless, some options clearly are better than others, so the goal maybe simply to find an optimal solution that can add value to the pre-existing state, drawn from a number of practical alternatives that can be identified during the design process. The creativity of design must be filtered by the pragmatics of implementation for a successful outcome to be possible. Even more so, the decision-making process must be robust, transparent and compatible with the appetite for the decision-makers to take risks. Optimisation is defined in this context as seeking the best mix of decisions from a range of identified choices that have a reasonable probability of success.

Chapter 9 explores the problem of identifying which adaptation opportunities offer the best chance of success. The focus in this chapter is adaptive reuse, how intervention potential can be modelled and how projects that are likely to succeed can be identified ready for further detailed enquiry. An integrated decision-making process applicable to individual building projects is outlined and demonstrated. It is shown that some building types have a greater likelihood of success than others, displaying different levels of uncertainty and risk exposure.

Chapter 10 looks at the technique of multiple criteria decision analysis in the context of evaluating sustainable development. It advocates that all building adaptation interventions can be assessed according to their value for money and quality of life contributions and explains the method for calculating these ratios and comparing the value add of competing alternatives. Sustainable development is an oxymoron, but the expected level of sustainability from a development can be modelled and used to make decisions that reflect the best combinations of objectives. It is argued that interventions that are feasible (measured in terms of value for money) and desirable (measured in terms of quality of life) can be integrated together to calculate a sustainability index capable of ranking and selecting any type of intervention for any type of building project.

Chapter 11 describes how the type of adaptation intervention can be determined and presented using a 3D spatial model at any point in a building's life cycle. The trigger for intervention is argued as a combination of physical condition (x), space utilisation (y) and triple bottom line reward (z). Where these attributes are strong projects are likely to be retained or extended, but where they are weak projects are likely to be disposed or reconstructed. Over time the status of a project will change and its 'xyz' coordinates will be capable of being mapped in 3D space where the linear distance from the project's coordinates to decision 'hotspots' can be objectively measured to determine the level of confidence involved in such decisions.

Chapter 12 concludes this part by reflecting on how buildings can be designed in the first instance to maximise successful adaptation later in their life cycle. A method for rating 'adaptivity' using a star-rating schema is discussed and defined. Its application to adaptive reuse is explored in particular through the proposition that high values of design adaptivity lead to high values of adaptive reuse potential later in life when the original building function becomes obsolete. It is also argued that ultimate success is still dependent on the choice of functional change as much as latent conditions that underpin an effective conversion, and making the appropriate decision in this regard is crucial.

Case studies are included throughout this part to help visualise the application of decision-making models. They comprise the Melbourne General Post Office (GPO), Bond University's Mirvac School of Sustainable Development (MSSD), 88 George Street Sydney, plus a diverse collection of completed adaptation projects from around the world.

9

Identifying Adaptive Reuse Potential

Adaptive reuse potential (ARP) describes the propensity of an asset to be 'recycled' to perform a significantly different function while keeping the basic attributes of the asset in place. In theory it applies to any asset. Adaptive reuse is distinguished from more general recycling in that, while it may be modified to suit a new purpose, it is not reprocessed to such an extent that its original form is lost. ARP can deliver sustainability outcomes through preserving materials that might otherwise be destroyed or sent to landfill and is a particularly efficient form of recycling because it involves reuse of the whole rather than just some or all of its subordinate parts. Where the extent of the reuse is high, the contribution to sustainability is more significant, and other benefits are likely to be secured. Adaptive reuse is of interest when the original function of the asset has become obsolete.

In the context of buildings, adaptive reuse will involve a change in function from its existing purpose to a significantly different one. An example might be to take an obsolete multi-floor industrial warehouse that otherwise has substantial physical life left in it and convert it into luxury residential apartments. The motivation for this intervention might be because the land upon which the project sits is in a good location (perhaps it occupies a river or harbour frontage in a central area of a major city), has appropriate access to public transport and other amenities, has heritage protection that forbids demolition and reconstruction, fulfils a shortage of high-end residential properties considered to be in demand and has the opportunity to reduce development costs through retaining a large proportion of the structure and building envelope.

Sustainable Building Adaptation: Innovations in Decision-Making, First Edition.
Sara J. Wilkinson, Hilde Remøy and Craig Langston.
© 2014 John Wiley & Sons, Ltd. Published 2014 by John Wiley & Sons, Ltd.

9.2   ARP Model

The conceptual framework for the ARP model was described in Langston et al. (2008). It has generic application to any countries and any existing building type. An estimate of the expected physical life of the building and the current age of the building, both reported in years, is required. In addition, an estimate of physical, economic, functional, technological, social, legal and political obsolescence has to be made. Obsolescence acts like an exchange rate to 'discount' the expected physical life of the building to arrive at its useful life. An algorithm in the model takes this information and produces an index of reuse potential as a percentage. Existing buildings in an organisation's portfolio, across a city or territory, can be ranked with this method according to the potential they offer for adaptive reuse at a particular point in their life cycle. When the current building age is close to and less than the expected useful life, the model identifies that planning for intervention should commence.

Buildings are like other assets – they deteriorate and eventually become obsolete as they grow old. A building's physical life, often interpreted as its structural adequacy or safety, is diminished by obsolescence, resulting in a useful life somewhat less than its notional physical life. Obsolescence may be defined as the loss of utility due to the development of improved or superior products or services in the marketplace, but not utility loss due to natural deterioration or decay. Nevertheless, accelerated decay arising from a lack of proper maintenance or renewal could be regarded as akin to physical obsolescence. Obsolescence, however, is also driven by other factors (Seeley 1983; Mansfield 2000; Douglas 2006). The useful (effective) life of a building in the past has been difficult to forecast due to premature obsolescence, arising from one or more of the following attributes:

- Physical
- Economic
- Functional
- Technological
- Social
- Legal
- Political

Some may argue that environmental obsolescence should also be included. In this work, environmental issues have been subsumed within the category of technological obsolescence. Furthermore, as the marketplace continues to become more sustainability conscious, social, legal and political obsolescence are likely to increasingly reflect the environmental agenda. So it is considered rather unhelpful to isolate environmental obsolescence as a discrete category in the ARP model.

Each obsolescence attribute needs to be quantified to determine the rate at which physical life is reduced per annum. Given obsolescence is not readily observable in the marketplace, a process of surrogate estimation is employed. A scale of 0–20% is used to assess obsolescence vulnerability, where 0% means it is effectively immune and 20% means it is significantly exposed, except in the case of political obsolescence where a scale of –20% to +20% is applied, such that –20% is seen as a supportive environment, 20% is seen as an inhibiting environment and 0% is apathetic. Interim scores in increments of 5% provide some level of discernment, while greater accuracy is considered unwarranted.

Useful life can be determined from Equation 9.1. The equation applies the notion that useful life is discounted physical life and therefore uses the long-established method of discount as its basis, where the 'discount rate' is taken as the sum of the obsolescence factors per annum:

$$\text{Useful life}(L_u) = \frac{L_p}{\left(1 + \sum_{i=1}^{7} o_i\right)^{L_p}} \tag{9.1}$$

where:

L_p = physical life (years)
O_1 = physical obsolescence (% as decimal per annum)
O_2 = economic obsolescence (% as decimal per annum)
O_3 = functional obsolescence (% as decimal per annum)
O_4 = technical obsolescence (% as decimal per annum)
O_5 = social obsolescence (% as decimal per annum)
O_6 = legal obsolescence (% as decimal per annum)
O_7 = political obsolescence (% as decimal per annum)

Using this approach, a building receiving the maximum reduction of 20% for each type of obsolescence will have a useful life calculated at about one-third of its notional physical life.

An index is computed that can be used to prioritise buildings according to their potential for adaptive reuse, expressing this potential as a percentage between 0 and 100. Buildings with a high index possess high potential, while buildings with a zero index have no potential. The ARP index provides a measure of benchmarking (identifying low, moderate or high potential for reuse in individual buildings), timing (understanding increasing or decreasing reuse potential and prioritising work) and ranking mutually exclusive projects (the higher the score, the more potential for reuse). The concept is illustrated graphically in Figure 9.1 (Langston 2008).

Values for EL_u (effective useful life), EL_b (effective building age) and EL_p (effective physical life) are determined by individually multiplying L_u, L_b and L_p by 100 and dividing by L_p. This enables a maximum scale for the x axes of 100 (i.e. 100% of the building's life cycle). L_b is defined as the

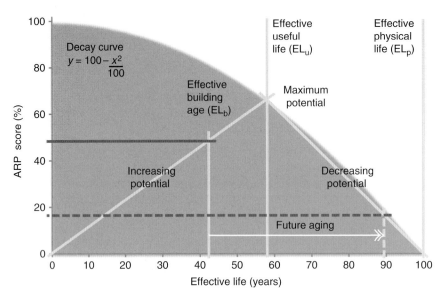

Figure 9.1   Langston's ARP model concept.

current age of the building (in years). The feasible zone for the ARP model is defined by the shaded area under the curve as defined by Equation 9.2:

$$y = 100 - \frac{x^2}{100} \qquad (9.2)$$

Equations 9.3 and 9.4 represent the lines of increasing and decreasing ARP, respectively:

$$\mathrm{ARP}_{(increasing)} = \frac{\left[100 - (EL_u^2/100)\right] \cdot EL_b}{EL_u} \qquad (9.3)$$

$$\mathrm{ARP}_{(decreasing)} = \frac{\left[100 - (EL_u^2/100)\right] \cdot (100 - EL_b)}{100 - EL_u} \qquad (9.4)$$

where:

EL$_u$ = effective useful life (years)
EL$_b$ = effective building age (years)

Values of ARP above 50 are considered to have high potential for adaptive reuse, while values between 20 and 50 show moderate potential, and values below 20 show low potential. An ARP value of 0 has no potential. Values above 85 would suggest strongly that planning activities should commence. When EL$_u$ and EL$_b$ coalesce, the maximum ARP value possible for that building is found.

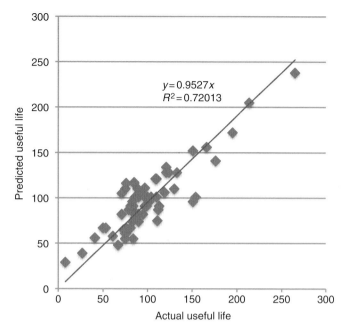

Figure 9.2   Correlation between ARP-predicted and actual useful life (n = 64).

Langston (2011a) found that the expected useful life derived from the ARP model and the actual useful life derived from 64 completed adaptive reuse projects worldwide were highly correlated. The line of best fit was computed as $y = 0.9527x$ (see Figure 9.2). In fact, if actual useful life was reduced by about 5% to account for inherent overestimation (e.g. ignoring the time taken to assess, document, approve and construct the proposed intervention strategy), the line of best fit would have been $y = x$, thus indicating a 45° line or perfect comparison. The degree of scatter was illustrated by an r^2 value of 0.72013, suggestive of a tight relationship. If the line of best fit was instead assumed to be $y = x$, then r^2 fell to just 0.69971. This is arguably a truer indication of reliability. Of course any correlation between predicted and actual useful life is normally illogical, but its use here demonstrates quantitatively the accuracy of a key part of the ARP method.

9.3   Obsolescence Rates

9.3.1   Physical Obsolescence

Physical obsolescence is an accelerated deterioration as a result of inadequate maintenance and repair regimes. It is therefore logical and plausible to look at budgeted maintenance expenditure over the life of a building, whether in hindsight or as an honest prediction. Records are normally kept to indicate this information, but it may need to be converted (updated) to the present in order to be interpreted. Annual budget allowances in real terms are then averaged.

Different buildings demand different levels of maintenance expenditure. As a way forward, a notional maintenance budget (excluding routine cleaning) should be calculated as a percentage of the updated initial construction cost against which current spending can be benchmarked. The appropriate percentage is considered generic to particular types of buildings based on their occupancy profiles. Assuming this percentage was 2% per annum, the notional budget for a building that has an updated construction cost of $10 million is $200,000. This amount would include the apportioned cost of salaried maintenance operatives, outsourced contracts, replacement materials and necessary equipment to facilitate the work.

Maintenance budgets must take account of significant replacement cycles, and therefore the amount in reality is unlikely to be the same each year. But if the average over many years is taken, a sense of the adequacy of resourcing can be obtained. Where data is not complete, estimation is required.

It is reasonable that physical obsolescence can be measured by an examination of maintenance policy and performance. Useful life is effectively reduced if building elements are not properly maintained. A scale is developed such that buildings with a high maintenance budget receive a 0% reduction, while buildings with a low maintenance budget receive a 20% reduction. Interim scores are also possible, with normal maintenance intensity receiving a 10% reduction.

So in the example in Section 9.3.1, $200,000 is considered normal and would result in a 10% reduction for physical obsolescence. Double that amount (or more) would receive a 0% reduction, while half that amount (or less) would receive a 20% reduction. Values for 5% and 15% reductions could be derived or interpolated accordingly. Further accuracy in computing the level of physical obsolescence is probably not warranted.

9.3.2   Economic Obsolescence

Economic considerations are often dominant in decisions concerning obsolescence in buildings. These relate fundamentally to ensuring that the income stream remains greater than the cost stream and indeed greater than other alternative opportunities of similar risk level. Failure to generate a regular operating surplus renders a building economically obsolete. Such obsolescence can offer advantage, however, as it instigates new investment in more productive and technically advanced infrastructure, which has higher income and hence higher operating surplus potential. The capital investment in delivering the new infrastructure is written off over many years and provides some residual value at the end of its economic life if it is on-sold.

It can be argued that relative price factors, and in particular the price of capital investment compared to labour in maintenance and repair activities, determine the speed with which capital goods become obsolete. A rise in real wages or other running costs, a reduction in the production price of capital works or a fall in the rate of interest will all tend to increase the rate of replacement investment and hence lower the average age of capital stock.

Economic obsolescence is affected by investment and yield decisions, not only for the building concerned but for other investment opportunities relevant to the property owner. While models exist to assess economic performance, an underpinning driver is business activity, which is a function of location and proximity to markets.

It is reasonable that economic obsolescence can be measured by the location of a building relative to a major city, central business district or other primary market or business hub. Useful life is effectively reduced if a building is located in a low-density demographic. A scale is developed such that buildings sited in an area of high population density receive a 0% reduction, while buildings sited in an area of low population density receive a 20% reduction. Interim scores are also possible, with average population density receiving a 10% reduction.

In the absence of specific population data per square metre of land area, it is practical to consider dense urban environments as having a 0% reduction, while rural environments receive a 20% reduction. Nevertheless, if city centres are devoid of people resident there after business hours, or personal safety is such that commerce and tourism are diminished, then locational heuristics might have to be reassessed to take account of where the people spend most of their time. With the continued expansion of e-business activities, proximity to communication and transport hubs might take on more significance.

9.3.3   Functional Obsolescence

Obsolescence is also a result of poor design and in particular design that cannot be readily changed or updated to meet new circumstances. Lack of flexibility of spatial layouts and difficulty in modifying services when change occurs lead to higher churn costs. Churn is defined as the disruption of space due to functional change, including the routine rearrangement of furniture when people, for example, move jobs or get promoted or when businesses expand or contract.

For businesses, direct churn costs are normally recorded, but indirect churn often goes unmeasured and unconsidered. For non-commercial buildings, churn might be better assessed in units of time rather than dollars. Modular design intended to support flexibility is likely to assist in the achievement of low churn costs and can be used to make assessments where other measures are unavailable.

Either way, it is reasonable that functional obsolescence can be measured by determining the extent of flexibility imbedded in a building's design. Useful life is effectively reduced if building layouts are inflexible to change. A scale is developed such that buildings with a low churn cost receive a 0% reduction, while buildings with a high churn cost receive a 20% reduction. Interim scores are also possible, with typical churn costs receiving a 10% reduction. While some subjectivity is involved here, making assessments in one of five categories (0, 5, 10, 15 or 20%) is not unwieldy.

9.3.4   Technological Obsolescence

Buildings are increasingly reliant on technology, and the rate of obsolescence of technology can be significantly higher that many other assets. Technology can comprise communication media, computer equipment, heating, ventilation and air conditioning plant, lighting, security, lifts, building management systems and other energy-dependent devices. So it is not unreasonable to consider energy usage as an indicator of technology intensity in buildings. Technological complexity may also be highlighted.

Energy is demanded to provide a comfortable environment that can additionally enhance occupant safety, productivity, satisfaction, health and personal empowerment. Low-energy systems, such as solar power, solar water heating, wind turbines, natural lighting and ventilation, tend also to utilise lower levels of technology, often in fact exploiting passive design solutions in preference to more mechanisation and complexity. Of all energy demanded for building operation, the main application is normally directed to provide for occupant comfort.

It is reasonable that technological obsolescence can be measured by the building's use of operational energy. Useful life is effectively reduced if a building is reliant on high levels of energy in order to provide occupant comfort. A scale is developed such that buildings with low energy demand receive a 0% reduction, while buildings with intense energy demand receive a 20% reduction. Interim scores are also possible, with conventional operating energy performance receiving a 10% reduction.

Depending on location and climate, buildings will demand different amounts of energy, and these amounts will further vary by season. Good data is now available in most countries to benchmark building performance in terms of operating energy. For example, if a 4-star energy rating (or its equivalent) is considered as demonstrating conventional performance, then the amount of energy demanded by such a building for a given climatic zone in a typical year can be used to draw conclusions about technological obsolescence in other similar-purpose buildings located in that zone. Where a building uses twice as much energy (or more) per square metre, a 20% reduction in obsolescence would be selected, and where a building uses half as much energy (or less) per square metre, a 0% reduction would apply. Using this approach, an estimate of technological obsolescence based on energy intensity can be reasonably determined.

Modern buildings tend to have significantly more technology embedded in them than older buildings. The ease with which this technology can be upgraded or changed also impacts on obsolescence. Expensive retrofits will make upgrade less attractive. This matter of 'serviceability' is considered part of the assessment of functional obsolescence.

9.3.5   Social Obsolescence

Changes in fashion and style can result in buildings becoming outdated even though the materials and technologies within them are still young. Building image in particular can drive market interest and enhance property values or rental income streams. Buildings that are intended as investments must be managed so as to better reflect market preferences and directions, and so it is logical to think that such buildings will have a greater propensity to become obsolete as a function of social change.

Market relevance can be affected by externalities such as a change in population demographics, consumer behaviour, fashion or style desires, locational status, image expectations, marketing intensity and other factors not directly related to building design or condition. Refurbishment to give a fresh look and to instigate change for its own sake is not uncommon for buildings that are highly reliant on market investment for their viability. Hence, the usage for which the building is designed will set up a profile of potential social obsolescence virtually from the day the building is completed.

It is reasonable to suggest that social obsolescence can be measured as the relationship between building function and the marketplace. Useful life for commercial buildings is effectively reduced if building feasibility is based on external income. A scale is developed such that buildings with fully owned and occupied space receive a 0% reduction, while buildings with fully rented space receive a 20% reduction. Interim scores are also possible, with balanced rent and ownership receiving a 10% reduction. Useful life for non-commercial buildings is effectively reduced if there is a decline in service demand or need. A scale is developed such that buildings with a strengthening demand or relevance receive a 0% reduction, while buildings with a weakening demand or relevance receive a 20% reduction. Interim scores are also possible, with a steady demand or relevance receiving a 10% reduction.

For commercial buildings, floor area dedicated to leasing expressed as a percentage of the total useable floor area can therefore be used to model social obsolescence. If 100% of usage is leased, then social obsolescence is maximised, and if 0% of usage is leased (i.e. no external income stream involved), then social obsolescence is minimised. Mixed developments can be suitably proportioned between these limits, and changes in this mix over time will have a corresponding effect on useful life.

For non-commercial applications (such as residential, religious, educational and civic buildings), measurable trends in demand or relevance can alternatively be used to model social obsolescence. Where data are not available, some judgment may need to be exercised. Buildings that serve a growing social purpose (i.e. needed for an increasing population) would be favoured over those where the purpose is declining or at risk (such as a decrease in religious observance) or otherwise dependent on current fads and fashion (such as boutique sporting and leisure facilities).

9.3.6   Legal Obsolescence

The design of buildings is governed by a plethora of standards, rules and approval processes that assign and confirm compliance. As time passes, there is a tendency for higher levels of compliance to be expected for new buildings, relatively making existing buildings more likely to require future upgrade. This upgrade often is triggered by substantial renovation or extension. Examples of the trend for increases in compliance can be seen in areas such as fire safety and egress provisions, disabled access requirements, smoke-free spaces, air circulation and cleaning protocols, toxic material avoidance and energy performance targets. New buildings are becoming more complex as a result of higher compliance levels, and older buildings are becoming obsolete faster as a result of their relative lower compliance.

Compliance is closely related to the quality of the original design and its translation via construction. High-quality design would suggest use of materials, components and systems that are above normal expectations, and in many cases these would deliver enhanced performance over cheaper alternatives. Higher quality also infers sophistication and the delivery of solutions that exceed minimum standards. For these reasons, legal obsolescence is tied to building quality standards.

It is reasonable that legal obsolescence can be measured by the quality of the original design. Useful life is effectively reduced if buildings are designed and constructed to a low standard. A scale is developed such that buildings of high quality receive a 0% reduction, while buildings of low quality receive a 20% reduction. Interim scores are also possible, with average quality receiving a 10% reduction.

But quality is a subjective matter and not easy to quantify. High-quality finishes, for example, do not necessarily mean that fire safety compliance is exceeded, and materials such as asbestos that were seen in their time as valid choices have turned out to be substantial liabilities. Nevertheless, the best indicator we have of building quality is price. By comparing the cost to construct per square metre (including design fees and fit-out but excluding land acquisition) across other buildings of similar type and location, it is possible to readily identify typical (normal) cost performance. High-quality buildings may be interpreted as costing a premium over normal cost in the order of 100% or more, while low-quality buildings may cost 50% or less. Other values can be interpolated accordingly. It would be best to compare construction cost in the context of when the building was constructed, rather than convert original costs to present-day equivalents, as the effect of price escalation over long periods of time can result in significant distortion. Where cost information is not available or relevant, more subjective decisions about building quality will be required.

9.3.7   Political Obsolescence

A few researchers have alluded to another type of obsolescence that arises through changes to zoning, ascribed heritage classification and other imposed regulatory changes. While this may be assumed as a subset of legal obsolescence, the approach adopted thus far is more focused on compliance with building standards. Particularly as heritage is an important aspect in building renovation, a new category of obsolescence is advocated with its own unique assessment.

It is reasonable that political obsolescence can be measured by the extent of public and local community interest surrounding a project. Useful life is effectively reduced if there is a high level of (restrictive) political interference expected. A scale is developed such that buildings with a low level of interest receive a 0% reduction, while buildings with a high level of interest receive a 20% reduction. Interim scores are also possible, with normal public and local community interest receiving a 10% reduction.

Where a project can receive a significant benefit from political interference, rather than a constraint, it is feasible to extend the assessment scores into the positive range (i.e. –20% to +20%). In this case, should the potential interference be seen as an advantage, it would extend a building's useful life and help offset other obsolescence considerations, which are all negative or neutral. An example of a positive influence would be government funding opportunities or enhanced tax concessions that can be accessed when pursuing an adaptive reuse strategy.

The advantage of this approach is to acknowledge formally that increased political interference will impact on the potential for adaptive reuse. Heritage protection is a particular example that is not otherwise considered. A heritage overlay may restrict opportunity for further development and limit what is possible or may boost tourism appeal through clever revitalisation of an otherwise neglected asset.

The disadvantage of this approach is the difficulty in assessing the level of public and local community interest and whether it is indeed positive or negative. Often this issue may be hidden and not arise until a particular media event or protest action occurs. Public buildings of note and/or buildings already with a significant heritage or environmental overlay should be carefully considered. Smaller private buildings free of redevelopment controversy should be assessed as neutral (i.e. a 0% obsolescence value).

Environmental obsolescence, rather than assessed in isolation, can now be more appropriately integrated with technological considerations (i.e. energy, performance and comfort), social considerations (i.e. consumer behavioural changes), legal considerations (i.e. sustainability contributions and compliance) and political considerations (i.e. mandated planning, conservation and regulatory frameworks).

9.4   Case Study: GPO Building, Melbourne

The *Melbourne GPO* is provided to illustrate the ARP model when applied in practice. The building was constructed on the corner of Elizabeth and Bourke Streets in 1859, following some earlier and modest structures dating back to 1837. In the following years up to 1867, a much grander two-storey building was developed. The building underwent further major renovations, which were completed in 1919, including a new sorting hall.

However, in 1992 Australia Post announced plans to end the GPO's role as a postal hub in favour of a number of decentralised mail centres. The building was effectively obsolete and planned to be sold. In 1993 a shopping centre was proposed, but indecision reigned and the permit lapsed. In 1997 a hotel was proposed, but this idea similarly did not proceed. Then again in early 2001, plans for a retail centre were announced and approved. The

Figure 9.3   Melbourne GPO interior (former sorting hall) after conversion.

project had to overcome a major setback when it was almost gutted by fire in September 2001. Nevertheless, the work finally took place and the building was opened to the public in October 2004.[1]

The building's conversion included a restoration of the main sorting hall (see Figure 9.3) and a contrasting modern extension to the northern end of the site (see Figure 9.4). Williams Boag Architects, Arup and the successful contractor St Hilliers worked closely together on this important project. It now stands as one of the more prominent and well-known adaptive reuse buildings in Australia. It subsequently won the RAIA National Award for Commercial Buildings and the Sir Osborn McCutcheon Commercial Architecture Award (2005).

Actual physical life of the project is unknown, despite its near demise in 2001 potentially providing the answer, and a cursory inspection would suggest there are many good years left. The physical life was estimated using a weighted model of factors comprising the building's environmental context, occupational profile and structural integrity (Langston 2011a). Using 1919 as the new base, the calculated life of 200 years suggested that the building would be structurally safe until 2119. The most recent renovations now extend this date well into the future, securing the building's heritage value to the City of Melbourne.

Obsolescence rates were assessed with the benefit of hindsight. Physical obsolescence was rated high since maintenance for much of the building's life was not a priority and as evidenced by accelerated deterioration that occurred. Economic obsolescence was zero as the building was in the centre

Figure 9.4   Melbourne GPO exterior (showing junction of new extension).

of Melbourne. Functional obsolescence was considered low as the building had substantial open space. The massive external walls of the building provided some thermal mass that would help insulate the interior from the outside conditions, but nevertheless some form of heating was essential, and the demand on energy would have been moderate. But from a social perspective, this building was owned and occupied by a government authority and did not rely on external income to survive. The building was constructed to a high standard and so legal obsolescence was low. It would be logical to assume that changes to the building would attract considerable community interest, so political obsolescence would be high as this may limit future redevelopment opportunities.

Therefore, obsolescence was assessed at 15, 0, 5, 10, 0, 5 and 20% respectively across the seven categories (total 55%), leading to an obsolescence rate over 200 years of 0.28% per annum. Applying Equation 9.1, useful life was calculated at 116 years. Given the reset base of 1919, the building would be expected to become obsolete in 2035. The reality was considerably less (a difference of 36.5%). Given these outcomes for physical and useful life, an ARP score of 49.1% was determined (see Figure 9.1). This index was interpreted as depicting moderate and increasing potential. A few years later and the building would have reached a score of 50% and be seen as high potential. The maximum ARP score possible was 66.7% given a useful life of 116 years (note this is 58% of the expected physical life, or 58 years on the 100-year scale in Figure 9.1). But substituting expected useful life with the actual useful life of 85 years, the ARP score would have risen

Physical life worksheet

Melbourne GPO   Suggested forecast (years) =   200

The building comprises a concrete structure and massive stone-faced masonry walls, steel roof framing with glass vaulted ceiling, large open plan atrium and perimeter offices.

y/n ?

Environmental context

Question		y/n
Is the building located within 1 kilometre of the coast?		n
Is the building site characterised by stable soil conditions?	#	y
Does the building site have low rainfall (<500mm annual average)?		y
Is the building constructed on a 'greenfield' site?		n
Is the building exposed to potential flood or wash-away conditions?		n
Is the building exposed to severe storm activity?		n
Is the building exposed to earthquake damage?		n
Is the building located in a bushfire zone?		n
Is the building located in an area of civil unrest?	#	n
Are animals or insects present that can damage the building fabric?	#	y

Occupational profile

Question		y/n
Is the building used mainly during normal working hours?	#	n
Are industrial type activities undertaken within the building?		n
Is the building open to the general public?		y
Does the building comprise tenant occupancy?		n
Is a building manager or caretaker usually present?	#	y
Is the building intended as a long-term asset?	#	y
Does the building support hazardous material storage or handling?		n
Is the building occupation density greater than 1 person per 10 m²?		n
Is the building protected by security surveillance?		n
Is the building fully insured?		y

Structural integrity

Question		y/n
Is the building design typified by elements of massive construction?		y
Is the main structure of the building significantly over designed?		y
Is the building structure complex or unconventional?		y
Are building components intended to be highly durable?	#	y
Are there other structures immediately adjacent to the building?		y
Was the workmanship standard for the project high?	#	y
Is the building founded on solid rock?		y
Is the roof susceptible to leaking in bad weather conditions?		n
Is the building protected against accidental fire events?		n
Is the building designed as a public monument or landmark?		y

Notes:
Questions indicated (#) are double weighted
Blank responses are ignored   100% completed

Adaptive reuse potential

Adaptive reuse potential (ARP%) =   **49.1**

The building was not well maintained during its life, and was partially destroyed by fire in 2001. But it is well located, with a large mail sorting atrium, is reasonably well insulated, owner occupied, well built and heritage listed.

Physical life (L_p) =	200 years	Index =	200
Building age (L_b) =	85 years	Override =	

Original construction date =	1859	Today's date =
Last refurbishment date =	1919	(enter only if refurbishment was major) 2004

Physical (O_1)	0.15
Economic (O_2)	
Functional (O_3)	0.05
Technological (O_4)	0.10
Social (O_5)	
Legal (O_6)	0.05
Political (O_7)	0.20
Total =	0.55

Obsolescence rate per annum =   0.28

Useful life (L_u) =	115.5 years	Adaptive reuse potential is moderate and increasing
Years to useful life =	30.5 years	$EL_u = EL_b$
Maximum ARP score (%) =	66.7	(assuming $L_u = L_b$)
ARP difference (%) =	35.9%	

Risk management:

Best case obsolescence =	0.45 (low)	
Useful life (L_u) =	127.6	
ARP% =	39.5	Adaptive reuse potential is moderate and increasing (no change)
Worst case obsolescence =	0.70 (high)	
Useful life (L_u) =	99.4	
ARP% =	64.3	Adaptive reuse potential is high and increasing
ARP difference (%) =	62.9	

Moderate

Notes:
It was only in recent years that the building became superfluous to the operations of Australia Post. At no time before that would the postal service be seen as in decline.

Figure 9.5 Melbourne GPO ARP worksheet (dated 2004).

to 81.9%. This makes a very strong case for adaptive reuse intervention on the basis of the substantial 'embedded physical life' still in the building.

The calculations embodied in the ARP model for the Melbourne GPO are summarised in Figure 9.5, including an assessment of risk exposure based on both optimistic and pessimistic obsolescence rates. There was only a moderate level of doubt over the assessment of obsolescence vulnerability, leading to a range of 0.45 to 0.70, so the expected value of 0.55 was close to the mean.

9.5   Discussion

Buildings are major assets and represent a significant financial investment. Although they are long lasting, they require continual maintenance and restoration over their life cycle. Buildings can become inappropriate for their original purpose due to obsolescence or can become redundant due to changes in demand for their services (Johnson 1996), and it is at these times that change is likely: commonly demolition to make way for new construction or some form of refurbishment or reuse (Langston and Lauge-Kristensen 2002).

Making better decisions about built assets can improve our performance from a sustainability perspective and deliver economic, social and environmental benefits to property owners, investors and other stakeholders. In particular, the reuse of valuable resources can offset the need to destroy existing buildings and contribute positively to climate change adaptation initiatives that are becoming increasingly urgent. An understanding of how long buildings last contributes to this discussion.

The ISO-15686 series on service life planning for buildings and constructed assets is a useful resource for assessing building durability. However, it is more applicable to individual systems and materials than the overall building as a single asset. The estimated service life of any component is calculated as its theoretical life expectancy multiplied by a series of factors each scored in the range 0.8 to 1.2 (i.e. 1 = no impact). The factors comprise (a) quality of components, (b) design level, (c) work execution level, (d) indoor environment, (e) outdoor environment, (f) usage conditions and (g) maintenance level. While a building is just a sum of the parts, these parts can be renewed and replaced, leaving the basic structure as the main determinate of overall life expectancy. Other literature on service life discusses the effect of external and internal actions on building durability and identifies location, usage and design as principal issues. This is underpinned by a large amount of technical data.

Obsolescence can be defined as the inability to satisfy increasing requirements or expectations (Iselin and Lemer 1993; Lemer 1996; Pinder and Wilkinson 2000). This is an area under considerable stress due to changing social demand (Kintrea 2007) and brings with it a raft of environmental consequences. Yet obsolescence should not mean defective performance. Douglas (2006) made the further distinction between redundancy and obsolescence – the former means 'surplus to requirements' – although this may be a consequence of obsolescence. Nutt et al. (1976b:6) took the view that '... any factor that tends, over time, to reduce the ability or effectiveness of a building to meet the demands of its occupants, relative to other buildings in its class, will contribute towards the obsolescence of that building'. Other researchers have identified political changes such as zoning, ascribed heritage classification and various regulatory interventions as an additional form of obsolescence (e.g. Luther 1988; Gardner 1993; Campbell 1996; Kincaid 2000).

Economic considerations, however, are often dominant in decisions concerning building obsolescence (Baum 1991). Barras and Clark (1996) argued that relative price factors, and in particular the price of capital investment compared to labour in repair and maintenance activities, determine the speed with which capital goods become obsolete. A rise in real wages or other running costs, a reduction in the production price of capital works, a fall in the rate of interest charged on borrowings and favourable changes to tax concessions will all tend to increase the rate of replacement investment and hence lower the average age of capital stock.

Haapio (2008) stated that reliable data for forecasting obsolescence are rarely available. Usually estimates are based on the experience and judgment of specialist consultants. Where products are replaced and discarded before their service life has been reached, the remaining service life is wasted. As Aikivuori (1996) attested in her study of housing refurbishment in the private sector, obsolescence-based refurbishment tends to occur earlier than deterioration-based refurbishment. Therefore, future obsolescence deserves more attention during design, including the benefits that long-life, loose-fit and low-energy characteristics deliver to society.

The concept of obsolescence appears not dissimilar to depreciation, but in the latter case monetary values are used rather than utility (or performance) to describe the effect. Depreciation is defined as a noncash expense that reduces the value of an asset as a result of wear and tear, age or obsolescence and typically involves setting aside money for replacement when its useful (effective) life has been reached. Depreciation is often calculated using a diminishing value method reflecting a negative exponential or decay curve.

Similar parallels can also be drawn to the technique of discounting, which reduces the value of an asset or cash flow today to take account of the real opportunity cost of money at some future time. Discounting also reflects a negative exponential curve. Depreciation and discounting share a common objective of measuring 'decay' in initial values.

The decline in value caused by obsolescence, just like in opportunity cost deliberations, is not necessarily a regular or fixed rate per year, but rather this is assumed in order to make the calculations more manageable in practice. The various types of obsolescence need to be considered, either by using the more dominant cause and ignoring others or by adopting their combined effect. It is likely, as is found with discounting, that the components of the rate can work in opposite directions, and therefore a stabilising (central tendency) effect is produced.

Langston (2011b) applied ARP and discounted physical life to the assessment of building 'archetypes' to test whether particular building typologies hypothetically had a better chance of achieving higher ARP scores when they became obsolete. The method adopted the Program Evaluation and Review Technique (PERT) used in planning and scheduling activities as a means of assessing the impact of task duration on overall completion of time estimates (Uher 2003). Ten different types of facilities were chosen for the study. PERT analysis was applied in a unique manner to assess the range of obsolescence values that could reasonably be expected for each facility classification. These were estimated from the experience of the author, from which the annual obsolescence rate, skew, obsolescence range, ARP score and coefficient of variation were computed.

The ARP model itself assumes that each obsolescence category is equally weighted. Subsequent research has found no evidence to warrant this assumption to be varied. The annual obsolescence rate, used to 'discount' physical life to useful life in the ARP model, is computed as the sum of obsolescence values across the seven categories, expressed as a fraction. The key results from Langston (2011b) are summarised in Table 9.1.

As explained earlier, ARP scores above 50% are described as 'high', scores between 20 and 50% are 'moderate', while scores below 20% are 'low'. All facility types except residential, industrial and religious displayed 'high' ARP scores.

This data can be expressed as an archetype to help visualise the impact of ARP for each facility classification. Archetypes are defined as patterns that have generic application. The derived archetypes are provided in Figure 9.6 (Langston 2011b). The higher the ARP score, the better is the potential of success. The shaded area indicates the likely range of ARP scores (i.e. large ranges are more uncertain). The solid triangle indicates the ARP profile, while the two dotted triangles indicate the range boundaries for best and worst ARP outcomes. A low skew value (i.e. <50%) indicates a more favourable ARP profile than a high skew value (i.e. >50%). The archetypes therefore provide key strategic advice at a glance.

It should be noted that prediction of physical life is not required when interpreting each archetype, as useful life is expressed as a percentage of physical life (i.e. life cycle %) rather than in years. However, years can be interpreted by making an explicit estimate of physical life. The number of

Table 9.1 Summary of archetype values for selected facilities.

	Annual obsolescence	Range	ARP score (%)	Life cycle (%)	Skew (%)	CoV (%)
Commercial (office tower)	0.51	0.65	63.7	60.2	20	58.5
Residential (detached house)	0.18	0.35	29.5	84.0	80	72.7
Retail (shopping centre)	0.79	0.15	79.3	45.5	33	4.1
Industrial (warehouse)	0.32	0.40	46.9	72.9	72	43.8
Landmark (museum)	0.65	0.30	72.6	52.3	50	11.6
Civic (community centre)	0.37	0.40	51.9	69.4	70	35.7
Recreational (hotel)	0.45	0.30	59.3	63.8	50	21.4
Healthcare (hospital)	0.73	0.35	76.4	48.6	18	12.9
Educational (school)	0.58	0.40	68.8	55.9	84	16.0
Religious (church)	0.25	0.60	39.3	77.9	86	78.0
Mean	0.48	0.39	58.8	63.0	56.3	35.5
CoV (%)	42.9	37.2	28.2	20.4	46.1	75.4

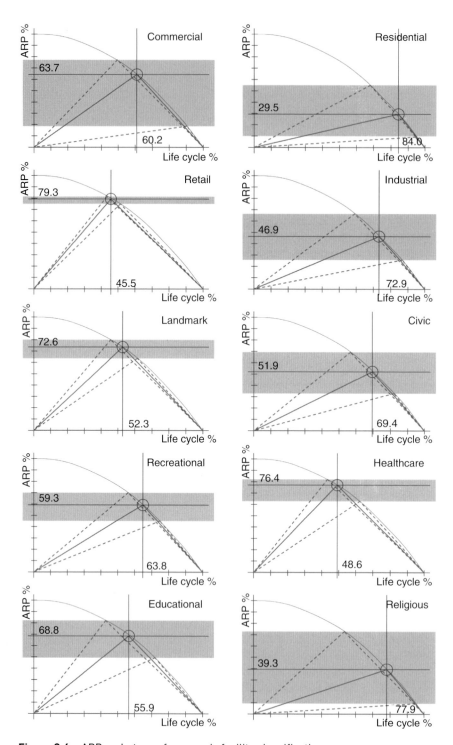

Figure 9.6   ARP archetypes for generic facility classifications.

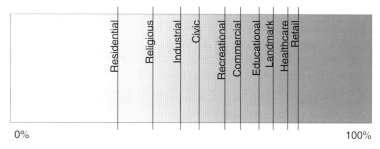

Figure 9.7   Expected ARP priority of generic facility classifications.

years of expected useful life, when added to the date of the original construction (or last major refurbishment), gives the date of optimum adaptive reuse intervention.

These results suggest that different facility classifications have different ARP profiles. The derived profiles are expected to apply to a majority of particular design solutions within each classification. Therefore, the facility classification is shown to be influential to the success of an adaptive reuse intervention and should be used as a key variable when searching for adaptive reuse opportunities. Based on likely estimates of obsolescence formulated via PERT analysis, facility classifications can be ranked, as shown in Figure 9.7.

Facility classifications such as retail, healthcare and landmark, for example, appear more attractive as potential adaptive reuse projects than facility classifications such as residential, religious and industrial. Other examined classifications fall somewhere between. Interestingly, retail, healthcare and landmark have lower uncertainty than the remainder, indicating a higher level of confidence in the prediction. Healthcare, commercial and retail have a low skew value, which makes them attractive since earlier intervention is likely, although commercial also has high uncertainty (i.e. large range).

The ARP model, by its nature, correlates ARP score with useful life and follows the decay curve as its regression line. The timing of adaptive reuse intervention is critical, for in order to maximise benefit, intervening too early or too late is counterproductive. The optimum intervention point, shown by the life cycle % value, has a mean across all facility classifications of 63% with a very low coefficient of variation. In other words, the optimum intervention point appears to be around two-thirds of the facility life cycle on average. The longest intervention period is for residential, which may partly explain why adaptive reuse to other functional purposes in this building type is rare. The time over which the benefit of the change can be enjoyed is therefore the shortest of all facility classifications examined.

9.6   Conclusion

The shape of the 'mountain' in Figure 9.1, depicting the rise and fall of ARP, is a function of the obsolescence factors that are deemed to apply. Rather than envisage these as accurate estimates, they are more appropriately understood as ranges within which reasonable estimates occur. High rates of obsolescence lead to lower useful lives and ARP profiles skewed towards the short term, while conversely low rates of obsolescence lead to higher useful lives and ARP profiles skewed towards the long term. The problem with the latter is that ARP scores are lower since the point of optimal intervention is delayed, leaving relatively little time to enjoy the benefits of the new purpose before the end of the facility's life cycle. Strategically, projects with the highest potential for adaptive reuse are those that have the greatest rate of premature obsolescence and ARP profiles skewed left.

Note

1   Further information can be found at http://www.melbournesgpo.com/#history.

References

Aikivuori, A. (1996) Periods and demand for private sector housing refurbishment, *Construction Management and Economics*, 14(1), 3–12.

Barras, R. and Clark, P. (1996) Obsolescence and performance in the Central London office market, *Journal of Property Valuation and Investment*, 14(4), 63–78.

Baum, A. (1991) *Property investment depreciation and obsolescence*, London: Routledge.

Campbell, J. (1996) Is your building a candidate for adaptive reuse? *Journal of Property Management*, 61(1), 26–29.

Douglas, J. (2006) *Building adaptation* (2nd edition), London: Elsevier.

Gardner, R. (1993) The opportunities and challenges posed by refurbishment, in proceedings of Building Science Forum of Australia, Sydney, Australia, Feb, 1993, pp. 1–13.

Haapio, A. (2008) Environmental assessment of buildings, PhD Dissertation, Helsinki University of Technology, Finland.

Iselin, D.G. and Lemer, A.C. eds. (1993) The fourth dimension in building: strategies for minimizing obsolescence, Committee on Facility Design to Minimize Premature Obsolescence, Building Research Board, Washington, DC, National Academy Press.

Johnson, A. (1996) Rehabilitation and re-use of existing buildings, in *Building maintenance and preservation: a guide to design and management* (2nd edition), Mills, E.D. ed., Oxford: Architectural Press, 209–230.

Kincaid, D. (2000) Adaptability potentials for buildings and infrastructure in sustainable cities, *Facilities*, 18(3), 155–161.

Kintrea, K. (2007) Housing aspirations and obsolescence: understanding the relationship, *Journal of Housing and the Built Environment*, 22, 321–338.

Langston, C. (2008) The sustainability implications of building adaptive reuse (keynote paper), in proceedings of CRIOCM2008, Beijing, China, October 31–November 1, 08, pp. 1–10.

Langston, C. (2011a) Estimating the useful life of buildings, in proceedings of AUBEA2011 Conference, Gold Coast, Australia, April 27–29, pp. 418–432.

Langston, C. (2011b) On archetypes and building adaptive reuse, in proceedings of PRRES2011 Conference, Gold Coast, Australia, January 16–19.

Langston, C. and Lauge-Kristensen, R. (2002) *Strategic management of built facilities*, Boston, MA: Butterworth-Heinemann.

Langston, C., Wong, F., Hui, E. and Shen L.Y. (2008) Strategic assessment of building adaptive reuse opportunities in Hong Kong, *Building and Environment*, 43(10), 1709–1718.

Lemer, A.C. (1996) Infrastructure obsolescence and design service life, *Journal of Infrastructure Systems*, 2(4), 153–161.

Luther J.P. (1988) Site and situation: the context of adaptive reuse, in *Adaptive reuse: issues and case studies in building preservation*, Austin, R. ed., New York: Van Nostrand Reinhold, 48–60.

Mansfield, J.R. (2000) The Cutting Edge 2000: much discussed, much misunderstood: a critical evaluation of the term 'obsolescence', RICS Research Foundation, London.

Nutt, B., Walker, B., Holliday, S. and Sears, D. (1976) *Housing obsolescence*, Hants: Saxon House.

Pinder, J. and Wilkinson, S.J. (2000) The Cutting Edge 2000: measuring the gap: a user based study of building obsolescence in office property, RICS Research Foundation, London.

Seeley, I.H. (1983) *Building economics: appraisal and control of building design cost and efficiency* (3rd edition), London: Macmillian Press.

Uher, T. (2003) *Planning and scheduling techniques*, Sydney: UNSW Press.

10

MCDA and Assessing Sustainability

10.1   Introduction

Multiple criteria decision analysis (MCDA) is a contemporary alternative to social cost–benefit analysis as a means of evaluating sustainable development. It avoids the problem of converting social and environmental performance into monetary terms simply so it can be combined with tangible costs and benefits and included in a discounted cash flow. Alternatively, MCDA enables performance measured in a range of appropriate units to be mixed into a single score to depict preferences.

Economic criteria are essentially expressed as a return on investment over a number of years. Benefit–cost ratio (BCR), defined as the sum of the discounted benefits divided by the sum of the discounted costs, determines this ratio. Values less than 1 suggest the proposed project should be rejected, while values greater than 1 are of interest (higher values are more attractive). For buildings, benefits represent income derived from their use, offset by costs incurred in earning that income. Such costs comprise capital expenditure such as construction and operating expenditure, such as cleaning, energy, maintenance and replacement, and are estimated as part of a (life-)cost planning process. Economic criteria, by their nature, are tangible.

Social criteria are often determined using a weighted matrix of criteria weights and performance scores that are multiplied together and summed. This approach is commonly employed in value management workshops to compare and rank alternatives. While still numerical, the assessment of social criteria is more subjective and lacks the rigour of economic evaluation. It may include any benefit that is not otherwise measured in monetary terms, such as aesthetics, flexibility, heritage preservation, sense of place, proximity and access and well-being. One of the issues that must be resolved

Sustainable Building Adaptation: Innovations in Decision-Making, First Edition.
Sara J. Wilkinson, Hilde Remøy and Craig Langston.
© 2014 John Wiley & Sons, Ltd. Published 2014 by John Wiley & Sons, Ltd.

is the relative weight of criteria, which is often assisted by the use of pair-wise comparison.

Environmental criteria lie somewhere in between economic and social criteria in terms of their assessment. While environmental impacts can be measured in a range of units like GJ/m^2, parts per million and kg CO_2e of carbon, the level of difficulty involved in their estimation is significant and time consuming. Life cycle assessment (LCA) and environmental impact assessment (EIA) are two common techniques to measure environmental consequences. Different issues will have different importance, and the level of uncertainty and risk will vary.

10.2   Background

There has been much research concerning the assessment of built environment sustainable development. Deakin et al. (2002) listed 61 assessment methods (as at September 2000) to underline this point. Included in this list were rating tools such as *BREEAM* and *LEED*; well-known evaluation techniques such as cost–benefit analysis, life cycle analysis, contingent valuation method, EIA and multi-criteria analysis; software solutions such as *BEPAC*, *ENVEST* and *INSURED*; and statistical methods such as analytic hierarchy process, concordance analysis and hedonic pricing. Some of these techniques are focused on environmental assessment, some include evaluation of wider social issues and some concentrate purely on monetary impacts.

Agenda 21 on Sustainable Construction (Bourdeau 1999) signalled to a new global context for evaluation. It identified cost, time and quality as the competitive factors in the traditional building process; then created a new paradigm by added resources, emissions and biodiversity; and then placed all this into the global context of economic constraints, social and cultural issues and environmental quality.

Guy and Kibert (1998:45) further discussed the local development of sustainability indicators in the context of Agenda 21, concluding that 'sustainability indicators are principally about awareness-raising and making environmental, economic and social sub-systems transparent to citizens and decision-makers'.

Cole (1999) discussed a range of issues including the difference between 'green' and 'sustainable', references and benchmarks, target performance levels, potential versus actual performance, qualitative and quantitative criteria and the use of weighting. He identified three primary dimensions of assessment: criteria (i.e. human, site, ecological), time (i.e. past, present, future) and scale (i.e. materials, components, site, community, regional, global). He concluded that the notion that a universally applicable tool would be widely adopted in different countries was highly questionable. Subsequently, Cole (2005) raised further doubt about whether existing methods are capable of being easily configured to fulfil what he called the emerging sustainability agenda. He suggested there is a move away from detailed performance measurement to encouragement of industry dialogue, understanding and take-up.

Li and Shen (2002) concluded that the uncertainty surrounding sustainability assessment demands the use of fuzzy-set theory. They identified that some data may be quantitative and readily extracted, but other data, particularly dealing with quality of life considerations, may involve a degree of vagueness and imprecision. The latter must rely on the subjective judgment of specialists and decision-makers. But guiding a decision-maker requires a systematic structuring of data. They advocated a balanced approach between 'crisp' and 'fuzzy' data.

Kaatz et al. (2005) advocated that stakeholder participation had a significant role in the assessment process. They found that;

'The main assumption underlying all sustainability assessments is the need to focus on a limited number of issues, identified as the most significant in a particular assessment context, without compromising on the comprehensiveness of the method. The use of a scoping procedure to narrow the scope of building assessment to the most significant issues will help highlight and effectively address problems that are relevant in the context of the developing world' (449).

In a later paper, Kaatz et al. (2006) raised further communication issues relating to building sustainability assessment – namely, integration (e.g. sustainable development principles, stakeholder values and stakeholder knowledge), transparency and accessibility (e.g. access to information and communication) and collaborative learning (e.g. transfer of knowledge and enhancing commitment and learning).

Lützkendorf and Lorenz (2006:334) found that 'existing design and assessment tools do not address the many economic, social and performance facets over the life span of a building, and do not provide building assessment results for all dimensions of sustainable development'. They summarised the maturity of sustainability evaluation as commencing with the assessment of technical building design and construction cost and, extending to life cycle costs (LCC) and environment impacts (LCA), the further introduction of social aspects and utility and an integrated model that evaluates technical building design in the context of economic, social and environmental criteria.

Turner (2006) posed the question as to how decision-makers can be offered a comprehensive yet coherent package of decision-making tools in order to facilitate the transition to sustainability thinking. Part of the answer perhaps lies in the difference between 'weak' and 'strong' sustainability. The former is founded on an economic investment rationale (e.g. see Constantino 2006), while the latter demands conservation of natural capital and the introduction of acceptability thresholds. Despite the balancing notion embedded in sustainable development, compromise is not always appropriate and may not please anyone.

Matar et al. (2008) discussed a 3D context for integrating sustainability into traditional construction practices. Their proposed framework comprised three dimensions: project life cycle phases, project executing entities and sustainability performance parameters. They employed a numerical approach to measure performance based on 18 criteria:

- Energy
- Land
- Water
- Materials
- Greenhouse gases (GHGs)
- Ozone
- Site ecology
- Solid waste
- Liquid effluents
- Noise
- Air quality
- Aesthetics
- Durability
- IEQ
- Adaptability
- Traffic
- Socioeconomic
- Culture

In each case they benchmarked project performance against a reference value, and their overall performance was assessed by the unweighted aggregation of project values compared to reference values. Results were presented visually via radar graphs. Matar et al. (2008) called for further research to be undertaken into the construction industry to develop and refine indices for each parameter and their relative weights.

Perhaps the final word goes to Atkinson (2008) who declared that sustainable development should be concerned with how current decisions about a portfolio of wealth, including both man-made and natural capital, impact upon future well-being. The challenge moving forward therefore is to find a means where the capital wealth and social utility can be balanced against resource consumption and environmental impact in a strong sustainability paradigm and to provide decision-makers with appropriate strategic advice to enable them to recognise and implement good outcomes.

10.3   A New Approach

10.3.1   Conceptual Framework

The underpinning literature makes it clear that built infrastructure decisions must be both feasible and desirable. This requires a combination of multiple objective and subjective criteria. But to be sustainable, decisions need to be expressed in the context of environmental consequences. While many refer to sustainable development as founded on the pillars of economic, social and environmental performance (i.e. John Elkington's triple bottom line[1]), suggesting that all three are positive and need to be maximised, this conflicts with the general proposition that sustainable development is a balance or compromise between progress and conservation and hence involves opposing ideologies.

Rather than treat environmental performance as another positive that should be maximised, there is merit in treating environmental loss as a negative that should be minimised. This is a fundamental revelation that has the crucial advantage of enabling the calculation of ratios to measure the degree of success between positive and negative outcomes. Both feasibility and desirability can be expressed as ratios of positive outputs (like investment return and functional performance) over negative inputs (like energy demand and habitat destruction). Value for money is a measure of feasibility and is defined as the ratio of wealth created compared to resources consumed. Quality of life is a measure of desirability and is defined as the ratio of delivered utility compared to ecosystem impact.

This approach was originally conceived by Langston and Ding (2001) and is consistent with the views of Lützkendorf and Lorenz (2006) mentioned earlier.

When expressed as ratios of investment output over environmental input, sustainability performance increases as these ratios are maximised. Sustainability therefore is not something that may be bolted onto conventional decision-making processes, but rather must be integral with it. This notion supports coordinate-based decision-making. If both x and y variables are determined on a standard scale for all built infrastructure decisions, then development strategies and options can be readily compared. Figure 10.1 illustrates the new conceptual framework for evaluating the sustainability of decisions using a classic quadrant model. The x-axis is quality of life (desirability), while the y-axis is value for money (feasibility). A z-axis is added to represent sustainability risk (r) on the knowledge that bigger developments have higher inherent risk than smaller developments – an issue that is otherwise overlooked in the calculation of ratios – as size is common to both numerator and denominator and cancels out. All three axes use a scale of 0 (low) to 5 (high).

This approach generates a set of 3D coordinates for each development choice within a $5 \times 5 \times 5$ cube. The calculated distance in 3D space from the decision coordinate to the high sustainability corner (or 'hotspot') in the model (i.e. 5, 5, 0) can be used to rank preferences. The shorter the distance, the more sustainable is the decision. While it is possible for x and y coordinates to exceed 5 due to high ratios (e.g. a very profitable low-carbon building), this is very unlikely and indeed probably due to ratios being overly optimistic or incorrect. Nevertheless, the distance could still be calculated once the decision-maker has been satisfied that it is a valid interpretation.

This model can clearly show the imbalance between economic and social preferences with cognisance to environment impact. This can be objectively computed as the right-angle distance between the x–y coordinate and the diagonal line shown from bottom left to upper right corners of Figure 10.1. The greater the distance, the more imbalanced is

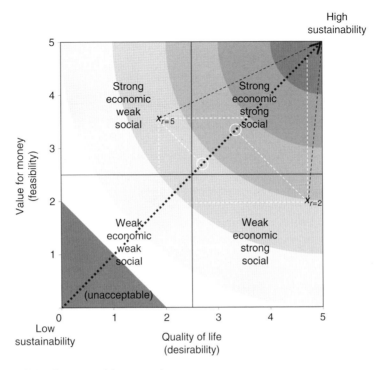

Figure 10.1 Conceptual framework.

the decision. High imbalance leads to a greater range of outcomes when bias between economic and social goals is considered. This range is shown by the vertical and horizontal dotted lines in the model and is a measure of stakeholder preference variability. The 3D coordinates objectively factor in the amount of imbalance between economic and social interests with the magnitude of the development in an economy and social context. Both increase the computed distance to the high sustainability hotspot of the model and hence have a negative impact on ranking and preference.

10.3.2 Value for Money

Value for money is a ratio made up of two parts. The numerator calculates net return with reference to a discounted cash flow of tangible costs and benefits over a mandatory 30-year time horizon. The denominator calculates resources consumed in the process of making that return, expressed in terms of embodied and operating energy inputs compared against legislated or customary targets over the same period. Equations 10.1, 10.2 and 10.3 are employed:

$$\text{value for money} = \frac{\text{return on investment}}{\text{energy usage}} \qquad (10.1)$$

where

$$\text{return on investment} = \frac{\sum \text{discounted benefits} \times 100}{\sum \text{discounted costs}} \qquad (10.2)$$

$$\text{energy usage} = \frac{\left[\text{actual embodied} + (\text{actual operating} \times 30)\right] \times 100}{\text{target embodied} + (\text{target operating} \times 30)} \qquad (10.3)$$

Return on investment is merely conventional BCR, defined as the ratio of time-adjusted benefits over costs as used commonly in capital budgeting decisions. It can be calculated in any currency in present-day terms adjusted by a real discount rate to reflect its timing. Energy usage is the sum of embodied energy (involved in initial construction and ongoing replacements) and operating energy (mainly electricity and gas), expressed in units of energy (commonly GJ or GJ/m^2). In these calculations a time horizon of 30 years is used to prohibit the opportunity for short-term decision-making, but cognisant of the fact that outcomes after 30 years are effectively negligible due to the compound effect of discounting.[2]

Value for money is based on potentially tangible data and forecasts and is robust and objective. It accounts for all project LCC and benefits adjusted for time equivalence (including capital works, cleaning, energy, maintenance and replacement together with the income stream generated by the project and its ultimate residual or scrap value) and all resource inputs consumed in that process (expressed in the form of detailed embodied and operating energy calculations), in both cases compared to an appropriate reference benchmark. The data used to develop Equation 10.1 can be as detailed as practicable. Referring back to the framework of Matar et al. (2008), value for money includes issues of 'feasibility' such as energy, materials, GHGs, durability and socioeconomic parameters.

10.3.3 Quality of Life

Quality of life is also a ratio made up of two parts. The numerator calculates utility (i.e. intangible reward) with reference to a scored list of weighted criteria. The denominator calculates impact with reference to a checklist of implemented environmental protection or mitigation measures such as pollution prevention. Equations 10.4, 10.5 and 10.6 are employed:

$$\text{quality of life} = \frac{\text{functional performance}}{\text{loss of habitat}} \qquad (10.4)$$

where

$$\text{functional performance} = \frac{\sum \text{weighted utility scores} \times 100}{\text{maximum possible score}} \qquad (10.5)$$

$$\text{loss of habitat} = \frac{\text{number of negative mitigation strategies} \times 100}{\text{total number of strategies in checklist}} \quad (10.6)$$

Functional performance uses a weighted matrix approach to assess non-financial performance on a scale of 0 (not applicable) to 5 (excellent) against a range of weighted criteria (e.g. aesthetics, comfort, safety). Loss of habitat is an assessment of the policies and procedures in place that are likely to minimise ecosystem damage and lead to better environmental outcomes and effectively estimate habitat risk. Both should be assessed in a non-judgmental equal-status multidisciplinary meeting of key stakeholders to limit bias and manipulation.

Quality of life is a robust and subjective ratio even though it is based on intangible data and forecasts. It accounts for all project nonmonetary contributions (including social and environmental utility) and all nonmonetary inputs used in that process (including loss of environmental resources and qualities), in each case compared to an appropriate reference benchmark. The latter can be informed by detailed LCA studies. Referring back to the framework of Matar et al. (2008), quality of life includes issues of 'desirability' such as land, water, ozone, site ecology, solid waste, liquid effluent, noise, air quality, aesthetics, IEQ, adaptability, traffic and culture parameters.

10.3.4   Sustainability Risk

Clearly value for money and quality of life ratios are independent of the scale of particular developments. For example, it would be inappropriate to compare a minor retrofit of a corner shop on an equal footing with construction of a low-carbon eco-city. While both may reflect sustainable development, the risk attached to delivering a successful eco-city compared to a successful retrofit is significantly greater. Sustainability principles also advocate 'less is more', suggesting that large developments are likely to be more pervasive and damaging to natural environments purely because of their footprint.

The level of risk could alternatively be calculated using a classic risk identification, analysis and response framework. But this level of complication is not warranted, given that risk is being used as a moderating factor on feasibility and desirability coordinates. A simpler scale is recommended as more practical, where 0 reflects no risk, 1 reflects minimal risk, 2 reflects moderate risk, 3 reflects significant risk, 4 reflects extensive risk and 5 reflects excessive (i.e. unacceptable) risk. As sustainability risk concerns potential 'influence', the construction cost or the development area can be used as tangible surrogates on which to base these judgments.

10.4   Life-Cost Planning

The determination of BCR is best achieved via a detailed examination of expected benefits and costs over an appropriate time horizon. In the latter case, costs can be estimated as part of a life-cost planning procedure. They comprise capital (e.g. land acquisition, purchase costs, construction, etc.) and operating (e.g. cleaning, energy, maintenance, replacement) expenditure. When allocated by year, a cash flow can be assembled ready for discounting and comparison. Given that BCR is always used to make decisions between one proposed project and others, the use of discounted cash flow analysis is essential. Up until this point, however, life-cost planning deals with constant dollars (i.e. present-day terms) and does not of itself involve discounted present value calculations.

Langston (2005) described cost planning as part of a cost control process that commences with the decision to build and concludes with the completion of design documentation. During this period the main objectives are:

- Setting cost targets, in the form of a budget estimate or feasibility study, as a framework for further investigation and comparison
- Identification and analysis of cost-effective options
- Achievement of a balanced and logical distribution of available funds between the various parts of the project
- Control of costs to ensure that funding limits are not exceeded and target objectives are satisfied
- Frequent communication of cost expectations in a standard and comparable format

The scope of cost planning should not be confined just to the construction of buildings, as is common in practice, but should include matters that are expected to arise during the life of the project. The process is aimed at improving value for money to the investor through comparison of alternatives that meet stated objectives and qualities at reduced expense. It would be inappropriate to ignore operational impacts.

The cost plan is one of the principal documents prepared during the initial stages of the cost control process. Costs, quantities and specification details are itemised by element or sub-element and collectively summarised. Measures of efficiency are calculated and used to assess the compliance of the developing design against budgetary assumptions. An elemental approach supports the comparison of individual building attributes with similar attributes in different buildings and forms a useful classification system.

Cost planning depends heavily on a technique known as 'cost modelling'. Modelling is defined as the symbolic representation of a system, expressing the content of that system in terms of the factors that influence its cost. Its objective is thus to 'simulate' a situation before it occurs in order that the problems posed will generate results that may be analysed and used to make informed decisions.

Life-cost planning is similar in concept to capital cost planning except for the types of costs that are taken into account and the need to express all costs in common dollars. The aim is to prepare a document that describes the composition of the building in a manner that is of use to the investor or owner in the longer term. A cost plan in this format could be used to demonstrate the relation between initial, replacement and running costs and to assist in the choice of durable specification and design details.

The objective of life-cost planning is not necessarily to reduce running costs, or even total costs, but rather to enable investors and building users to know how to obtain value for money by knowing what these costs are likely to be and whether the performance obtained warrants higher levels of expenditure. A project is cost effective if its life costs are lower than those of alternative courses of action for achieving the same objective.

Expenditure commonly associated with commercial buildings includes acquisition, cleaning, energy, rates and charges, maintenance and replacement. These costs can apply to the building structure, its finishes, fitments, services and external works. Plant and equipment, in particular, require specialist technical advice on performance statistics and energy demands. The establishment of life costs may be difficult without historical data or expert knowledge.

The life-cost plan should be prepared on at least an elemental basis showing all quantities and unit rates and should be set out in a fashion that enables analysis of totals for each type of cost category, including capital cost, along with costs/m^2 and percentages of total building cost. Of course life-cost planning, like any other form of cost investigation, is most effective in the early stages of design.

10.5   Case Study: Bond University Mirvac School of Sustainable Development (MSSD) Building, Gold Coast

10.5.1   Method

A case study methodology is employed here to examine the new approach explained in section 10.3. Case studies are often employed by researchers to illustrate the relevance of results in real-world settings (e.g. Ding 2005). While the objective of a decision model is normally to compare development options to identify and rank preferences, in this instance the selected case study is used to test the practicality of its deployment in industry. Validation of the underpinning model can be found in Ding (2004). Some of the details of the model have been improved, but essentially the content of the underpinning model is unaffected.

The case study chosen is the Mirvac School of Sustainable Development (MSSD) building at Bond University, Gold Coast, Australia. This building is the Australia's first 6-star *Green Star* educational building and represents high sustainability performance in the context of the *Green Star* methodology (which is similar in concept to *LEED*, *BREEAM* and other green rating tools). It would be expected that the case study lies in the upper right-hand quadrant of the model designated as 'strong economic strong social' development performance.

The MSSD building won many awards for its design and innovation, including the prestigious RICS Sustainability Award in 2009. The building has an area of approximately 2500 m^2 spread over three levels and is relatively small compared to other campus buildings. It was opened in October 2008 and now houses approximately 30 staff offices together with common area, teaching space and laboratories (see Figure 10.2).

Figure 10.2   MSSD building, Gold Coast.

The choice of an existing case study is made here so that the benefit of hindsight from over 3 years of occupation can help inform its measurement. Cost information is expressed in 2008 terms when the building was finished, although inflation to the current date would have no impact on ratios (i.e. BCR is unaffected by inflation). Designed performance is used unless actual performance to the contrary is available. Data are sourced from a combination of internal documents, public domain information and expert opinion.

Expected improvements in worker productivity are often cited in relation to sustainable buildings and sometimes are factored directly into cash flows as though they were tangible. This is not introduced into the case study for two reasons. First, there is considerable doubt over the longevity let alone the legitimacy of productivity improvements in green buildings despite their common inclusion as pseudo income to help justify higher upfront expenditure. Second, productivity improvements were never identified as a motivation for the construction of the MSSD building, and therefore it was considered inappropriate to include this criterion as part of the functional performance index either.

10.5.2   Return on Investment

The first problem that arises is that there is no income stream for the building. It is owned and operated by the university for the purposes of carrying out its educational business. However, the cost of leasing similar quality space in the local community (i.e. Varsity Lakes) has been taken as an indicator of the potential of the building to earn income. Further income is added for advertising rights on the building's exterior. The construction cost is well documented, but the operating costs are forecast based on a past audit of performance. All costs and benefits are expressed as $/m² of gross floor area. A real discount rate of 3% per annum is applied. Table 10.1 shows the cash flow that underpins the calculation of BCR.

The return on investment index using Equation 10.2 is computed as 181.23.

Table 10.1   Return on investment calculations.

Year	Real cost (A$/m²)	3% discounted cost cash flow	Real benefit (A$/m²)	3% discounted benefit cash flow
0 (initial)	5200	5200	—	—
1	50	49	495	481
2	50	47	495	467
3	50	46	495	453
4	50	44	495	440
5	100	86	495	427
6	50	42	495	415
7	50	41	495	402
8	50	39	495	391
9	50	38	495	379
10	200	149	495	368
11	50	36	495	358
12	50	35	495	347
13	50	34	495	337
14	50	33	495	327
15	100	64	495	318
16	50	31	495	308
17	50	30	495	299
18	50	29	495	291
19	50	29	495	282
20	200	111	495	274
21	50	27	495	266
22	50	26	495	258
23	50	25	495	251
24	50	25	495	244
25	100	48	495	236
26	50	23	495	230
27	50	23	495	223
28	50	22	495	216
29	50	21	495	210
30	200	82	495	204
30 (residual)	—	—	5200	2142
		6536		**11,845**

Table 10.2   Energy usage calculations.

	Actual building performance (GJ/m^2)	Normal performance or target (GJ/m^2)
Embodied energy (initial + replacement)	18.0	24.0
Operating energy (annual × 30 years)	4.4	11.9
	22.4	**35.9**

10.5.3   Energy Usage

The second problem that arises is that no embodied energy calculation was formally done on this building. However, based on a separate comprehensive study of commercial buildings in Melbourne, it is estimated that the embodied energy of a similar building is about 18 GJ/m^2 (Langston and Langston 2007). Given this figure takes account of the significant recycled content in the MSSD building, a typical target figure for this class of building would be closer to 24 GJ/m^2. The actual operating energy is 40.74 kwh/m^2 per annum. The target figure should be based on a 4-star *Green Star* performance level that reflects accepted current practice, which according to the Green Building Council of Australia is equivalent to 110 kwh/m^2. Table 10.2 shows the working that underpins the calculation of energy usage.

The energy usage index using Equation 10.3 is computed as 62.45. The resultant value for money ratio using Equation 10.1 is therefore 2.92.

10.5.4   Functional Performance

While energy performance was obviously a dominant driver for the building in order to meet its 6-star *Green Star* objective, it was not the only performance criterion. Some criteria related directly to its function as a high-standard university facility. When listed out it is clear that not all criteria have equal importance, so a weighting has been applied using a scale of 1 (low importance) to 10 (essential). Issues that are measured objectively, like operating energy levels and durability, are excluded here. Each criterion is rated for its perceived success using a scale of 0 (not applicable) to 5 (excellent). The assessment of utility is made using on-site experience and observation and discussions with colleagues. Table 10.3 shows the working that underpins the calculation of functional performance.

The functional performance index using Equation 10.5 is computed as 89.82.

Table 10.3 Functional performance calculations.

Performance criteria	Weighting (1–10)	Performance score (0–5)	Weighted score	Maximum score possible
Aesthetics	6	3	18	30
Flexibility	5	4	20	25
Market image	10	5	50	50
Occupant comfort	8	5	40	40
Socialisation	6	5	30	30
Living laboratory	8	5	40	40
Security	3	3	9	15
Proximity to transport	4	5	20	20
Innovation	5	4	20	25
			247	**275**

10.5.5 Loss of Habitat

A checklist is assembled of unweighted procurement strategies likely to impact on natural ecosystems. Some of these attributes are positive (upside risk), and some are negative (downside risk). Developments that reflect a mostly positive list of strategies are assessed as having a low risk of habitat damage, and vice versa. All factors used in the model are considered to have equal risk influence for the purposes of practical assessment and have been chosen accordingly, although a weighting system could be introduced if a suitable method for determining relativities were to be devised. Where questions give rise to both positive and negative responses, the dominant response is selected. The total list of strategies equals 33, and 23 of the assessed responses are considered positive (i.e. the potential risk impact is in the moderate range of 20–39%). Table 10.4 shows the working that underpins the calculation of loss of habitat.

The loss of habitat index using Equation 10.6 is computed as 27.27. The resultant quality of life ratio using Equation 10.4 is therefore 3.29.

10.5.6 Sustainability Index

The x coordinate for the model is 3.29 and the y coordinate is 2.90, which suggests the case study is very balanced between economic and social preferences (i.e. distance to the diagonal line in the model is just 0.28) and what would be expected given a triple bottom line philosophy was applied by the design team. The building is therefore placed in the 'strong economic strong social' quadrant as anticipated. The z coordinate is assessed as $r = 1$, given that the building is quite small in area. The resultant distance from these coordinates to the high sustainability hotspot is 2.89, and this calculation would be used as a comparative measure against other options or developments.

If the diagonal line in the model from lower left to upper right is expressed on a scale of 5 (where 5 represents 'high sustainability'), this case study

Table 10.4 Loss of habitat calculations.

Strategy	Yes/no
Manufacture	
Does the manufacturer have an environmental management plan?	N
Are new raw materials a renewable resource?	Y
Does the manufacturing process involve hazardous materials?	N
During manufacture, are GHG emissions minimal?	Y
Does the manufacturing process generate untreated pollution?	N
Are product components manufactured from recycled materials?	Y
Are the majority of raw materials imported from overseas?	N
Is manufacturing waste sent to landfill?	Y
Are significant amounts of manufacturing waste recycled?	Y
Are most products packaged?	Y
Design	
Is environmental performance a specific design objective?	Y
Were outcomes evaluated using a life-cost approach?	N
Was embodied energy considered in the decision process?	Y
Are there significant heritage implications to be considered?	N
Construction	
Will the construction process generate untreated pollution?	N
Will environmental impacts during construction be monitored?	Y
Will construction waste be primarily recycled?	Y
Usage	
Does the intended function use water efficiently?	Y
Will pollutants be discharged directly into the environment?	N
Is waste recycled?	Y
Are significant energy minimisation strategies in place?	Y
Is noise transmitted to surrounding spaces?	Y
Demolition	
Are most demolished materials recyclable?	Y
Does nonrecyclable waste involve hazardous materials?	N
Are all components sent to landfill biodegradable?	N
Has a deconstruction plan been developed?	N
Context	
Is the site in a remote location?	N
Is the site environmentally sensitive or protected?	N
Was an environmental impact statement prepared for the project?	N
Are there rare or endangered species near the site?	N
Will the site's natural features be significantly disturbed?	Y
Is site stability and erosion control a particular objective?	Y
Are affected site areas reinstated upon completion of construction?	Y

would score 3.10 assuming an equal balance between economic and social preferences, with a very small variation range. These values are computed from the model and may be referred to as the 'sustainability index' (SI) for the development. This index, however, becomes a function of the balance preference or bias selected (e.g. 50:50, 25:75, 100:0) and ignores the development's scale and influence, forming the third dimension of the decision.

An SI less than 1 indicates the development is unacceptable and should be discarded. It is also possible to introduce additional thresholds that must be achieved, such as BCR must be ≥1, actual energy must be less than target energy (unless the differential is obtained from renewable sources), the weighted utility score must be at least half of the maximum possible score and the risk to habitat must be less than 50% of the maximum risk score. These thresholds can only be met by rebalancing the mix between return on investment and the other core criteria, or in other words, by spending more money upfront in order to secure better ongoing performance.

10.6   Discussion

Terms such as 'strategic assessment', 'triple bottom line accounting' and 'integrated assessment' are not well understood in practice and applied in an inconsistent manner in the literature (Hacking and Guthrie 2008). There are major difficulties with the assessment of sustainability, and the plethora of conceptualisations and terminology has only led to an equally diverse range of techniques and methods used to appraise sustainable development (Dietz and Neumayer 2007; Bond and Morrison-Saunders 2011). Indeed, Gasparatos et al. (2008) explored these issues and concluded there is no singular agreed definition and no singular agreed approach to assessment. Yao et al. (2011) highlighted this as a significant barrier to the evaluation of sustainable infrastructure, especially at the conceptual or design phase of development when key decisions are made.

Lozano (2008) noted that sustainability is inherently a difficult concept for people to understand. He advocated, like others, that visual representation methods be used to help communicate results. Visual MCDA offers great application to sustainable development including decisions concerning sustainable infrastructure. It is argued that modelling choices in 3D space enable decision preference to be determined by measuring the linear distance between project coordinates and optimum performance hotspots. This distance can be used to measure and rank options.

Decision-making needs to be more transparent and open. Mascarenhas et al. (2010) discussed the benefits of addressing sustainable development at the local scale, citing boundary, scale and knowledge efficiencies as examples. At the organisational or project level, the fostering of technological innovation is often considered to be an important element of broader sustainable development policy initiatives. Kain and Söderberg (2008) advocated public participation in decision-making as a potential solution. Scenarios can contribute to general debate or establish foundations for future policy discussion (Larsen and Gunnarsson-Östling 2009). Assessment approaches need to encourage stakeholder participation and engagement, but this objective is often thwarted by complexity and excessive technicality.

Morrissey et al. (2012) argued that actors in the built environment are increasingly considering environmental and social issues for development projects alongside functional and economic targets. Infrastructure projects represent major investment and construction initiatives comprising attendant environmental, economic and societal impacts across multiple levels. To date, while sustainability strategies and frameworks have focused on wider national aspirations and strategic objectives (Fiala 2008), they are noticeably weak in addressing micro-level decision-making in the built environment (de Meester et al. 2009), particularly for infrastructure projects.

There is a need to refine the decision-making process for assessing sustainability concerning built environment projects so that decisions are relatively easy to enact, understand and visualise by stakeholders and community leaders. This involves integrating the various aspects of sustainability rather than dealing with discrete elements of the problem. Lützkendorf and Lorenz (2006:355) defined the problem as 'balancing economic and social development with environmental protection'. This is a very insightful definition and was the motivation for the model presented here.

The case study demonstrates clearly that the model is both workable and practical. The model can also support more detailed calculations than those described should this be warranted or if information is available. While every attempt has been made to use authentic data, functional performance and loss of habitat are by their nature subjective and open to interpretation and bias. A software application called *SINDEX* was developed to help perform these calculations, albeit involving fairly simple maths. Nevertheless, the contrast between 'crisp' and 'fuzzy' data and the level of falsification that is possible in the latter case are acknowledged.

It is somewhat surprising that the derived x and y coordinates are not larger, especially given this building has iconic status in Australia as the first of its type to secure 6-star *Green Star* accreditation. But further reflection on this point suggests the reasons why. While it is true that the building is high performance green, as evidenced by the low energy and habitat values, and works well in providing a high level of utility to its occupants, it was expensive to construct at over double the price of a normal building of this type. Adjusting the construction price and residual value to $2600/m^2$ (i.e. 50% of the construction cost and an average price for new office space) leads to a value for money ratio of 4.38, a distance to the high sustainability hotspot of 2.08 and an imbalance measure of 0.77. The SI increases from 3.10 to 3.84, and the potential variation range is extended accordingly. The extra cost is attributed to the innovative nature of the building. There is no evidence of improvements in worker productivity due to 'green' design to defend higher procurement costs either.

The case study does not pretend to offer any generalisations for other built facilities, but merely presents a method for assessing alternative design choices from a broad sustainability perspective. Decisions to proceed with a development are often predicated on the understanding that the actual performance upon completion will reflect that originally anticipated, but this cannot be guaranteed. There are many examples of failed sustainability projects. So an interesting feature of the model is its ability to compare designed performance with actual performance as part of a post-occupancy evaluation. In this research, however, a single measure of performance was constructed with the benefit of hindsight.

It may be reasonably argued that the 3D model is overly simplistic. Indeed, how can 60+ rating tools, evaluation techniques, software solutions and statistical methods be replaced with a single decision model based on a couple of ratios and a moderating variable for scale? But the simplicity is deceptive. For example, Equation 10.2 (return on investment) comprises all capital and operating costs and benefits for a project over a 30-year time horizon, expressed as discounted present value, related to the building owner or occupier. Sitting behind this seemingly straightforward cash flow lies a detailed life-cost plan, embedding decisions about maintenance regime, component lifespans, climatic conditions, occupancy profile and economic factors such as funding costs, inflation and depreciation. Equation 10.3 (energy usage) comprises the corresponding carbon footprint for a project over a 30-year time horizon. Sitting behind this calculation are all the upstream energy intensities for building materials, allowance for recycled content, construction methods and usage patterns, as well as a comparison between energy performance and expected or normal practice. High-performance green buildings that are energy efficient are evaluated in relation to the costs of their acquisition and ownership.

Equations 10.5 and 10.6 that underpin the quality of life ratio are less factual. Nevertheless, rigour can be added to their evaluation by involving stakeholder participation and expertise and detailed investigations such as LCA or EIA performed by independent experts. The determination of weightings for utility can be assessed using pairwise comparisons and sensitivity analysis employed to test their impact on derived coordinates. Weights are applied to performance measures to reflect project objectives, but not to mitigation strategies as these are considered to have equal risk influence. Different facility types, however, may demand different lists.

MCDA is pivotal to the design of the model. There are four primary criteria; each is measured in different units, which are combined to determine the x and y coordinates needed to calculate the distance to the high sustainability hotspot and the value of the SI. Existing tools often deploy a single unit of measure and try to convert criteria into that measure so that a single value can be produced. Social cost-benefit analysis, used in environmental economics to convert nonmonetary criteria into discounted costs and benefits, is one such example. It bases decisions for environmental goods and services on capital budgeting criteria such as net present value, BCR and internal rate of return and is now largely discredited.

A key question is whether the new model is better than existing models that have been variously used for built environment sustainability assessment over many years. This question cannot be easily answered, as there is no accepted basis to make the evaluation. The advantage of the model lies in the breadth of its coverage and the visualisation of its results. The disadvantage of the model lies in the focus on a single facility, not a precinct or portfolio of assets, and determination of an appropriate mix between value for money and quality of life coordinates. The model can be improved by visualising risk analysis through identification of best- and worst-case scenarios for each coordinate and drawing a sphere centred on the original coordinate that embraces the other two outcomes. In this case risk is computed by assessing the volume of the sphere (i.e. small volume equals low risk, big volume equals high risk).

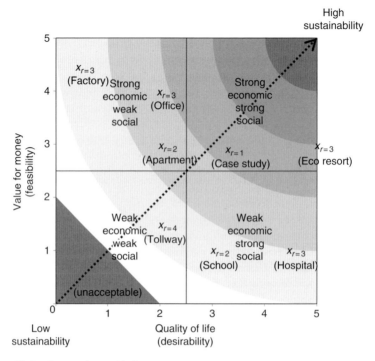

Figure 10.3   Comparison of infrastructure types.

As Lützkendorf and Lorenz (2006:352) pointed out, existing assessment methods 'must evolve if they are to add further value and address the array of emerging complex needs [... and] tools limited to environmental considerations will not meet the requirements of sustainable development currently emerging in the property and construction sector'. Coordinate-based decision-making based on feasibility (value for money) and desirability (quality of life) criteria, each expressed in the context of environmental impact and moderated for scale, is advocated as an appropriate method for model integration and visualisation.

How would the case study compare with other types of built infrastructure? Is the MSSD building competitive as an example of sustainability? To answer these questions, a large number of assumptions were applied to seven diverse and hypothetical developments. Figure 10.3 plots the outcomes. Table 10.5 presents the x, y and z coordinates for each development and ranks them according to the 3D distance to the high sustainability hotspot. It is clear that the MSSD case study is the most sustainable of the options in this list.

When considering the SI, even assuming a 50:50 balance between economic and social preferences, a slightly different rank list is produced. For example, the ecotourism resort has a much higher SI than the MSSD case study as this indicator ignores the scale of this development and its associated potential risks. This underlies the difference between the calculated 3D distance and the SI and is the reason why the former should be preferred when ranking options. The inclusion of development risk is an important

Table 10.5 Ranking of infrastructure types.

Development type	x y z coordinate			Imbalance distance	2D* distance	3D* distance	Sustainability index (SI)	Rank
MSSD case study	3.29	2.90	1.00	0.28	2.71	2.89	3.10	1
Ecotourism resort	5.00	3.00	3.00	1.41	2.00	3.61	4.00	2
Residential apartment	2.00	3.00	2.00	0.71	3.61	4.12	2.50	3
High-rise office Tower	2.00	4.00	3.00	1.41	3.16	4.36	3.00	4
High school	3.00	1.00	2.00	1.41	4.47	4.90	2.00	5
Public hospital	4.50	1.00	3.00	2.47	4.03	5.02	2.75	6
Chemical factory	0.50	4.50	3.00	2.83	4.53	5.43	2.50	7
Tollway	2.00	1.50	4.00	0.35	4.61	6.10	1.75	8

*Measured to the high sustainability 'hotspot'.

consideration when sustainability and precautionary (risk averse) principles are given due respect. This aspect of the model could be further enhanced through an algorithm to measure development scale more precisely.

10.7 Conclusion

The approach for evaluating built environment projects described in this chapter is applicable equally to new development or adaptation of existing buildings. The concept can be extended to precincts comprising multiple buildings by assessing each separately and calculating an average SI or by amalgamating inputs as though buildings were a single project.

Adaptation projects would often have lower construction costs and higher operating costs, especially maintenance and replacement, but could deliver savings on energy by integrating environmental performance measures into the refurbishment. Depending on the extent of reuse, significant savings can also be expected in embodied energy. When combined with other benefits like cultural and heritage protection, better use of existing infrastructure, less demolition waste being sent to landfill and minimisation of ecological footprint through the reclamation of carbon embodied in existing materials, adaptation projects are likely to be more sustainable than new developments.

Notes

1   See Cannibals with forks: the triple bottom line of twenty first century business, Capstone Publishing, Oxford 1997, although the term was first coined by John in 1994 (http://www.johnelkington.com/pubs-books-business.htm).

2   It may be argued that energy should also be discounted since it is likely that future delivery and usage patterns for energy will be more efficient than technology today can provide, and this can be easily incorporated into the model if deemed necessary. Embodied carbon or CO_2 equivalent GHG emissions (kg CO_2e or kg CO_2e/ m^2) could also be used as a substitute for energy.

References

Atkinson, G. (2008) Sustainability, the capital approach and the built environment, *Building Research and Information*, 36(3), 241–247.

Bond, A.J. and Morrison-Saunders, A. (2011) Re-evaluating sustainability assessment: aligning the vision and the practice, *Environmental Impact Assessment Review*, 31(1), 1–7.

Bourdeau, L. (1999) *Agenda 21 on Sustainable Construction*. CIB Report Publication 237, International Council for Research and Innovation in Building and Construction, Rotterdam: CIB.

Cole, R. (1999) Building environmental assessment methods: clarifying intentions, *Building Research and Information*, 27(4–5), 230–246.

Cole, R. (2005) Building environmental assessment methods: redefining intentions and roles, *Building Research and Information*, 35(5), 455–467.

Constantino, N. (2006) The contribution of Ranko Bon to the debate on sustainable construction, *Construction Management and Economics*, 24(7), 705–709.

Deakin, M., Huovila, P., Rao, S., Sunikka, M. and Vreeker, R. (2002) The assessment of sustainable urban development, *Building Research and Information*, 30(2), 95–108.

Dietz, S. and Neumayer, E. (2007) Weak and strong sustainability in the SEEA: concepts and measurement, *Ecological Economics*, 61(4), 617–626.

Ding, G.K.C. (2004) The development of a multi-criteria approach for the measurement of sustainable performance for built projects and facilities, PhD Thesis, University of Technology, Sydney, Australia.

Ding, G.K.C. (2005) Developing a multicriteria approach for the measurement of sustainable performance, *Building Research and Information*, 33(1), 3–16.

Fiala, N. (2008) Measuring sustainability: why the ecological footprint is bad economics and bad environmental science, *Ecological Economics*, 67(4), 519–525.

Gasparatos, A., El-Haram, M. and Horner, M. (2008) A critical review of reductionist approaches for assessing the progress towards sustainability, *Environmental Impact Assessment Review*, 28(4–5), 286–311.

Guy, G.B. and Kibert, C.J. (1998) Developing indicators of sustainability: US experience, *Building Research and Information*, 26(1), 39–45.

Hacking, T. and Guthrie, P. (2008) A framework for clarifying the meaning of triple bottom-line, integrated, and sustainability assessment. *Environmental Impact Assessment Review*, 28(2–3), 73–89.

Kaatz, E., Root, D.S. and Bowen, P.A. (2005) Broadening project participation through a modified building sustainability assessment, *Building Research and Information*, 33(5), 441–454.

Kaatz, E., Root, D.S., Bowen, P.A. and Hill, R.C. (2006) Advancing key outcomes of sustainability building assessment, *Building Research and Information*, 34(4), 308–320.

Kain, J-H. and Söderberg, H. (2008) Management of complex knowledge in planning for sustainable development: the use of multi-criteria decision aids, *Environmental Impact Assessment Review*, 28(1), 7–21.

Langston, C. (2005) *Life-cost approach to building evaluation*, Amsterdam: Elsevier.

Langston, C. and Ding, G. (2001) *Sustainable practices in the built environment* (2nd edition) Boston, MA: Butterworth-Heinemann.

Langston, Y. and Langston, C. (2007) Building energy and cost performance: an analysis of thirty Melbourne case studies, *Australasian Journal of Construction Economics and Building*, 7(1), 1–18.

Larsen, K. and Gunnarsson-Östling, U. (2009) Climate change scenarios and citizen-participation: mitigation and adaptation perspectives in constructing sustainable futures, *Habitat International*, 33(3), 260–266.

Li, H. and Shen, Q.P. (2002) Supporting the decision-making process for sustainable housing, *Construction Management and Economics*, 20(5), 387–390.

Lozano, R. (2008) Envisioning sustainability three-dimensionally, *Journal of Cleaner Production*, 16, 1838–1846.

Lützkendorf, T. and Lorenz, D.P. (2006) Using an integrated performance approach in building assessment tools, *Building Research and Information*, 34(4), 334–356.

Mascarenhas, A., Coelho, P., Subtil, E. and Ramos, T.B. (2010) The role of common local indicators in regional sustainability assessment, *Ecological Indicators*, 10(3), 646–656.

Matar, M.M., Georgy, M.E. and Ibrahim, M.E. (2008), Sustainable construction management: introduction of the operational context space (OCS), *Construction Management and Economics*, 26(3), 261–275.

de Meester, B., Dewulf, J., Verbeke, S., Janssens, A. and Van Langenhove, H. (2009) Exergetic life-cycle assessment (ELCA) for resource consumption evaluation in the built environment, *Building and Environment*, 44(1), 11–17.

Morrissey, J., Iyer-Raniga, U., McLaughlin, P. and Mills, A. (2012) A strategic project appraisal framework for ecologically sustainable urban infrastructure, *Environmental Impact Assessment Review*, 33(1), 55–65.

Turner, R.K. (2006) Sustainability auditing and assessment challenges, *Building Research and Information*, 34(3), 197–200.

Yao, H., Shen, L.Y., Tan, Y.T. and Hao, J.L. (2011) Simulating the impacts of policy scenarios on the sustainability performance of infrastructure projects, *Automation in Construction*, 20(8), 1060–1069.

11

Modelling Building Performance Using *iconCUR*

11.1   Introduction

Existing facilities require constant attention, maintenance and adaptation over their lives, which can be extensive. Facilities managers must make decisions such as whether to retain, dispose, extend, refurbish or adapt buildings in order to protect their value and usefulness. A tension exists between the inevitable effects of decay, including obsolescence, and the investment of new capital necessary to maintain performance. What strategy might be appropriate, what actions to take, when to intervene and whether it is worthwhile are all fundamental questions that need to be answered correctly.

A model has been developed, known as *iconCUR*, to help answer these types of questions. It is a 3D spatial model that uses multiple criteria to visualise the performance of an existing built asset at any time during its life cycle. The key criteria are physical condition, space utilisation and triple bottom line reward representing x, y and z coordinates that are plotted within the boundaries of a cube. Vertical edges of the cube represent optimal intervention decisions, and the distance from the coordinates at any point in time to the closest upper corner (or 'hotspot') informs the decision-maker about what to do.

One key benefit of *iconCUR* is that it produces a single asset performance value that combines a large number of considerations across a diverse spectrum. This is the essence of MCDA. Given that this value is on a fixed scale (0–5), properties can be compared within a portfolio to determine relative priority. Furthermore, the score can be used to judge current performance against threshold service expectations over time. The asset performance score can be used to produce the standard asset management profile so frequently referred to in the literature, yet do so with actual data rather than just a theoretical concept. These values when

Sustainable Building Adaptation: Innovations in Decision-Making, First Edition.
Sara J. Wilkinson, Hilde Remøy and Craig Langston.
© 2014 John Wiley & Sons, Ltd. Published 2014 by John Wiley & Sons, Ltd.

viewed retrospectively over time describe the property's performance history and may be colour coded for easy interpretation.

The combination of MCDA and tools to assist decision-makers to visualise the problem and proposed solution lies at the cutting edge of contemporary asset management. Interestingly, the model's conceptual framework, criteria, attributes and assessment are generic for any physical asset (i.e. whether investment, public or personal property) other than unimproved land, and it is only when defining the model's elements and their performance scores that the model becomes specific to built facilities (i.e. commercial, residential, industrial).

11.2   Visual MCDA

MCDA is a term applied to a collection of formal approaches for group or individual decision-making that incorporates multiple criteria (Belton and Stewart 2002). It has been applied to sustainable development problems in disciplines as diverse as land-use planning (e.g. Stewart et al. 2004), forest management (e.g. Laukkanen et al. 2004), wetland protection (e.g. Herath 2004), wildlife management (e.g. Berbel and Zamora 1995), mining (e.g. Martin et al. 1996), transportation (e.g. Mergias et al. 2007), portfolio management (e.g. Subbu et al. 2007), fisheries (e.g. Mardle and Pascoe 2002) and built environment assets (e.g. Kaklauskas et al. 2005).

More specifically, MCDA has been applied to mainstream construction and property problems, such as how to select appropriate refurbishment opportunities, as well as a range of wider asset management applications. Different refurbishment applications are reported in the literature, ranging from simple decision models (e.g. Kincaid 2002) to complex ones (e.g. Bostenaru Dan 2004). There appears no industry consensus on what key criteria, attributes and weighting should be adopted, as these are dependent on the particular context and, at least for weightings, stakeholder preferences. Sun et al. (2008) noted that despite the rising attention on asset management decision-making in more recent times, few practical publications in this area exist. Value management is a well-accepted technique within the built environment and is a particular instance of MCDA, although other evidence of acceptance or implementation of MCDA tools in practice is scarce.

Contemporary MCDA discussions frequently deal with issues of sustainability. Langston (2005) described a decision-making model called *SINDEX* that is used to choose between competing building designs. It was based on four key criteria, namely, maximising wealth, maximising utility, minimising resources and minimising impact, each given weights according to economic or social preferences. These criteria were combined as a 'sustainability index' where the higher the index, the more sustainable the alternative and where an index less than 1 meant the project was unacceptable. Threshold benchmarks also need to be met for each criterion. Langston (2012a) extended this work to incorporate the use of coordinate geometry in visualising decisions. This research was summarised earlier in Chapter 10.

Ribeiro and Videira (2008:113) highlighted that 'building renovation decision-making is a complex process that involves many stakeholders and relies on multi-dimensional information […] and can hardly be made efficiently if the appropriate decision-making aids are not employed'. Several authors have studied refurbishment decision-making in recent years, such as Rosenfeld and Shohet (1999), Douglas (2006), Alanne (2004), Gann et al. (2003) and Georgiou (2008). Rosenfeld and Shohet (1999) identified that the state of deterioration found in building components and systems largely determines their need for renovation. Douglas (2006) confirmed that changing expectations and utilisation patterns were seen as influential in decisions to adaptively reuse existing buildings through altering their primary function. Alanne (2004) proposed a multi-criteria 'knapsack' of techniques to identify and select feasible refurbishment actions during the conceptual phase of urban renewal projects. Gann et al. (2003) used multiple criteria to specifically model the quality of building design. Georgiou (2008) highlighted the need to predict future events in refurbishment decision-making, often in the absence of any real evidence of final success.

There have been others who have applied various forms of MCDA to solve more general real estate problems in recent years (e.g. Tupenaite et al. 2008; Aguilar 2009; Urbanaviciene et al. 2009; Yau 2009; Cebi and Kahraman 2010; Aznar et al. 2010; Roper et al. 2010; Rosato et al. 2010; Vadrevu et al. 2010; Ensslin et al. 2011). An evolutionary process appears to have taken place, beginning decades ago with quite narrow and cost-based evaluation tools to today's more complex and often visual models that merge multiple quantitative and qualitative attributes into a single decision criterion. This evolution continues, with ideas drawn from a wide range of knowledge fields comprising science and anthropology through to business and communications.

However, despite the diversity of attempts, there remains a knowledge gap concerning how to make rigorous yet practical decisions about existing built assets. In particular, as pleaded by Henig and Buchanan (1996), the need for mapping criteria, attributes and alternatives remains critical. Trinkhaus and Hanne (2003) proposed an interesting multiple criteria support system called *KnowCube* that comprised the three 'dimensions' of knowledge: organisation, generation and navigation. It originated from a focus on non-expert user needs and the requirement to visualise and interact with a decision problem. This work had roots in earlier research by Mareschal and Brans (1988) who invented the *g*eometrical *a*nalysis for *i*nteractive *a*ssistance (GAIA) procedure that is now commonplace in generic MCDA visualisation software[1] and identifies a possible path for MCDA development into the future.

Langston and Smith (2012) subsequently proposed a visual MCDA model called *iconCUR* for making better decisions about existing built assets. It is a modern example of MCDA with particular reference to the range of property management interventions that may be appropriate at various stages throughout a facility's life cycle. Just like *KnowCube* and *SINDEX* that preceded it, *iconCUR* is based on three key criteria capable of visual representation in 3D space.

11.3   *iconCUR* Model

The *iconCUR* model uses three primary criteria of *c*ondition, *u*tilisation and *r*eward to map the current status of a built facility in 3D spatial terms at any point in time during its life cycle. The x and y coordinates identify appropriate property management decisions, while the z coordinate indicates the strength of those decisions in terms of value add.

In simple terms, the following criteria and proposed actions lie at the heart of the *iconCUR* model:

- *Low condition and low utilisation*: reconstruct or dispose
- *High condition and high utilisation*: retain or extend
- *Low condition and high utilisation*: renovate or preserve
- *High condition and low utilisation*: reuse or adapt

Condition (x-axis) and utilisation (y-axis) are both measured using a scale of 0 (low) to 5 (high). The various relationships are neither good nor bad but merely reflect the actions that may be appropriate at various points in a property's life. The identified action may, for a range of reasons, not be worth the effort, so reward (z-axis) is intended to quantify whether such intervention is feasible. Reward is defined in the model as collective utility and comprises a combination of financial, social and environmental benefits similarly measured on a scale of 0 (low) to 5 (high). Any positive differential that results between expected reward before and after intervention measures the theoretical 'value add' from the proposed decision. However, high existing reward suggests there is little opportunity for improvement and vice versa. Collective utility is intended to represent the overall net benefit to stakeholders and overlooks the fact that there are potentially individual winners and losers from any decision. In other words, collective utility is concerned that the benefits to winners outweigh the costs to losers. It is complicated by the fact that not all project stakeholders have equal standing and engagement, so the reward values are moderated by the relative standing of individual stakeholders to which they accrue.

The model takes on the shape of a cube. Each vertical edge indicates an optimum decision outcome (from low to high reward), and spatial coordinates within the cube describe a property's current performance. Two key influences are at work: decay and restitution. These can be illustrated by an upward sloping plane drawn from the reconstruct/dispose corner representing low reward (i.e. 0, 0, 0) to the retain/extend corner representing high reward (i.e. 5, 5, 5). Properties deteriorate over time and therefore exhibit loss of condition, utilisation and reward as they succumb to natural decay. Property managers can resist this trend by injecting new money to upgrade their facility, and therefore condition, utilisation and reward may be enhanced as they invest capital. A 'push–pull' effect results, so that where a lot of investment occurs, the property rises up the plane towards the upper retain/extend corner (i.e. becomes a more important asset) and where little investment occurs, the property slides down the plane towards the lower reconstruct/dispose corner (i.e. becomes a less important asset). In the latter case, redevelopment opportunities emerge for the current owner or for a future owner.

Interim actions also exist. These comprise opportunities for property managers to retrofit (i.e. low condition, moderate utilisation), recycle (i.e. low utilisation, moderate condition), refresh (i.e. high condition, moderate utilisation) and repair (i.e. high utilisation, moderate condition) their facilities.

Finally, moderate condition and moderate utilisation suggest minor reconfiguration to the property in a variety of directions may occur, or just a watching brief be maintained. A 2D representation of the *iconCUR* model is provided in Figure 11.1 (the third dimension of reward is not shown here for clarity). The scale for all axes can be interpreted generally as 0–1 (very low), 1–2 (low), 2–3 (moderate), 3–4 (high) and 4–5 (very high).

The distance in space between the current mapped coordinates of a property and the corners and edges of the 3D cube can be used to rank properties within a portfolio of assets. For example, properties closest to the reuse/adapt edge (when measured horizontally to it) indicate the degree of certainty that is attached to this type of intervention decision. A low current reward suggests an opportunity to add value exists. Should such a decision be implemented, the position of the property in 3D space will then change, ideally to increase the value of all three coordinates and move it closer to the retain/extend corner representing high reward. The greater the distance between the old and new coordinates, the greater the impact of the decision, and provided the values rise, the greater the expected success.

Over the full life cycle of a property, a 'trail' will be left within the 3D space that describes the various interventions that occurred. This trail may comprise decay curves tracking downwards, interrupted by sudden leaps upwards as remedial actions are taken. Such results can be graphed in a 2D

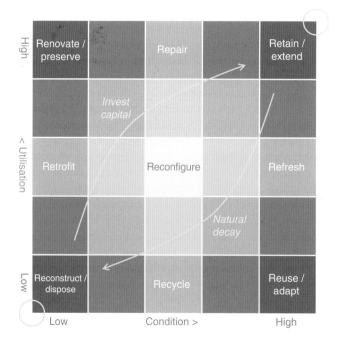

Figure 11.1 *iconCUR* model framework.

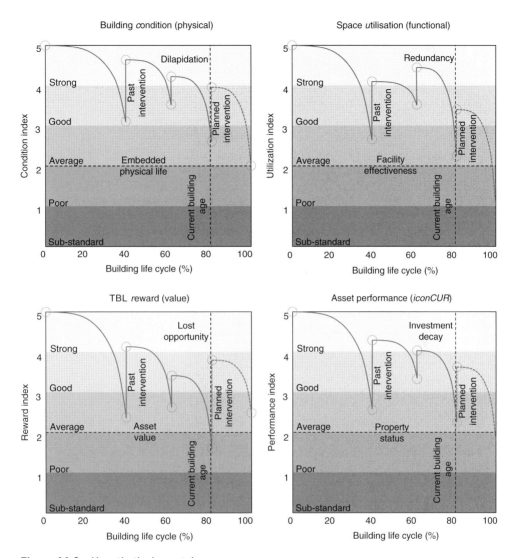

Figure 11.2 Hypothetical asset decay curves.

format to understand the property's status (i.e. asset performance) over time as demonstrated in Figure 11.2. Overall asset performance can be calculated as the average of the x, y and z coordinates in a particular time period.

11.4   Case Study: 88 George Street, Sydney

11.4.1   Overview

The case study selected in this chapter is the former Bushells Tea Company building (see Figure 11.3) located at 88 George Street in the historic 'Rocks' precinct of Sydney, Australia. Two scenarios are examined, representing the

Figure 11.3   88 George Street, Sydney (2010).

status of the building before and after major intervention works that took place in 2008. The building started its life as two separate industrial warehouses constructed in 1886 and 1912, further extended in 1986 and then adaptively reused to modern commercial office space in 2008.

The case study represents Australia's first example of a heritage-listed adaptive reuse project awarded a 5-star *Green Star* rating by the Green Building Council of Australia. A 5-star rating is approximately equivalent to Gold *LEED* certification in the US. This building remains the only known example of a successful heritage-listed green adaptive reuse conversion[2] completed in Australia at the time of writing.

Information for this study was obtained from fieldwork and enquiries (primary data) and websites, brochures and magazine articles (secondary data). The Sydney Harbour Foreshore Authority is the owner and manager of the building. The design team comprised Terroir (project architect), Steensen Varming (mechanical/electrical engineer), Hooker Cockram (contractor), Design 5 (heritage), Warren Smith and Partners (hydraulic engineer), Chris Bylett and Associates (quantity surveyor), Simpson Design Associates (structural engineer), Trevor Howse (fire engineer), Acoustic Studio (acoustics) and Access Australia (DDA consultant).

The Sydney Harbour Foreshore Authority set out to make this project Australia's most sustainable heritage-listed building by combining excellence in green design with an innovative approach to air conditioning using a future district cooling system. It has now set a new benchmark for heritage refurbishments by both government and the private sector. The project delivers approximately 2200 m^2 of commercial office and retail space over six levels, plus basement. All office levels enjoy excellent natural light and harbour views to the north and east, incorporating the Sydney Opera House and the northern Sydney CBD skyline.

Its key environmental features, making it an excellent example of what has been classified as 'green' adaptive reuse (Langston 2011c), comprise:

- *Indoor air quality*: the air conditioning system provides ventilation levels that are 50% higher than the Australian standard and provide tenants with a very comfortable working environment; functioning windows allow tenants to turn off the air conditioning and take advantage of ocean airflow and external temperature conditions.
- *Natural light*: large external windows on most facades allow tenants to enjoy high levels of natural daylight; high-quality and efficient lighting balances tenant comfort and energy use; more than 60% of office space has an external view, with no point more than 12 m from a window.
- *Thermal mass*: the thermal mass of the existing sandstone walls and floor helps stabilise internal temperatures; the use of insulation reduces internal noise levels and improves occupant comfort; the building's original heritage fixtures, including its facade and structure, have been conserved.
- *Choice of materials*: low volatile organic compound (VOC) paints, adhesives and sealants and products with low or no formaldehyde; polished timber and natural flooring in common areas promotes occupant well-being and health; timber either is recycled or comes from sources certified.
- *Energy strategies*: district-based air conditioning that delivers energy savings of up to 40% and prevents approximately 136,000 kg CO_2 from entering the Earth's atmosphere each year, individual metering on every floor and lighting zones of less than 100 m^2 on every floor.
- *Transport*: fewer parking spaces installed to encourage tenants to travel to and from work by cycling, walking or using public transport; cycling facilities with easily accessible showers, change facilities and secure storage; and a number of parking spaces are designated for small cars only.
- *Water*: expected 85% reduction in potable water use, replacing the existing water-cooled air conditioning with efficient VRV air conditioning using the future district cooling system, installing high-efficiency dual flush toilets and 6-star tap and sink fittings and individual water metering.
- *Emissions*: using the harbour to exchange heat from the building's air conditioning system, minimising facade lighting and positioning it to prevent light spillage and the use of non-ozone-depleting refrigerants in the cooling systems.

11.4.2   Before Intervention

The *iconCUR* model, applied retrospectively prior to the latest redevelopment decision (see Figure 11.4), suggests that adaptive reuse was appropriate, with a condition rating of 3.55 (out of 5) and a utilisation rating of 0.96 (out of 5), placing the project closer to the reuse/adapt edge of the cube than any other. A reward rating of 2.28 (out of 5) suggests opportunity remains

iconCUR

Property management matrix

Condition : Utilisation : Reward

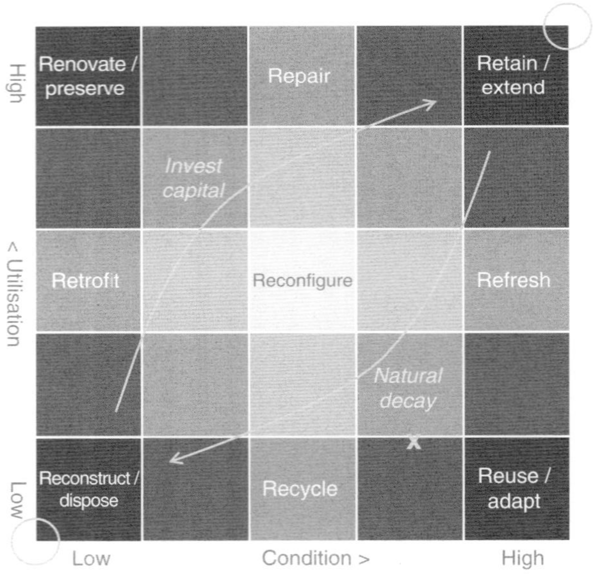

Matrix grid labels:

	Renovate / preserve	Repair	Retain / extend
High	*Invest capital*		Refresh
< Utilisation	Retrofit	Reconfigure	*Natural decay*
	Reconstruct / dispose	Recycle	Reuse / adapt
Low	Low	Condition >	High

iconCUR Assessment worksheet

88 George Street, Sydney (Australia)
Original construction: 1886 — 126 years old — (before proposed intervention)

Condition 3.55

	Design standard 50%	Maintained service level 25%	Regulatory/ compliance 25%	(high) 100%
Weighting				
Structure 20%	4	4	4	4.00
Exterior envelope 30%	4	4	4	4.00
Interior finishes/fitout 20%	3	3	4	3.25
Engineering services 20%	2	3	3	2.50
External works 10%	4	3	5	4.00
100%	3.40	3.50	3.90	3.90

Utilisation 0.96

	Demand or relevance 40%	Fitness for purpose 40%	User satisfaction 20%	(very low) 100%
Weighting				
Internal space (feca) 30%	1	2	2	1.60
External space (uca) 10%	0	2	0	0.80
Outdoor site area 10%	0	0	0	0.00
Equipment and fitout 30%	1	1	0	0.80
Engineering systems 20%	1	1	0	0.80
100%	0.80	1.30	0.60	0.60

Collective utility 3.00

	Economic performance 20%	Culture and heritage 40%	Environmental values 40%	(high) 100%
Weighting				
Operational viability 30%	2	2	3	2.40
Locational context 20%	5	5	1	3.40
Risk and opportunity 20%	2	4	3	3.20
Asset valuation 20%	3	5	1	3.00
Profile/mission 10%	4	5	2	3.60
100%	3.00	3.90	2.10	2.10

Stakeholder interest 3.80

	Short-term perspective 50%	Medium-term perspective 25%	Long-term perspective 25%	(high) 100%
Weighting				
Building owner 40%	3	4	5	4.20
Building user 20%	3	2	5	1.80
Facility manager 10%	5	5	1	5.00
Sponsor/financier 10%	5	3	1	2.60
Community 20%	5	5	5	5.00
100%	3.80	3.80	3.80	3.80

Reward 2.28

	Collective utility	Stakeholder interest	(moderate)
	3.00		

Asset performance 2.26

	Condition (x axis)	Utilisation (y axis)	Reward (z axis)	(average)
Property status	3.55	0.96	2.28	Recycle
Trending decision	5.00	0.00	0.00	Reuse/adapt
Decision exposure				Moderate risk
Proximity to corner		75%	67%	Far
Extra reward likelihood				Moderate

iconCUR that change is recommended and worthy of further investigation

iconCUR model 30/10/2010

22/09/12 16:31

ARC Linkage Project

Figure 11.4 *iconCUR model output (before intervention).*

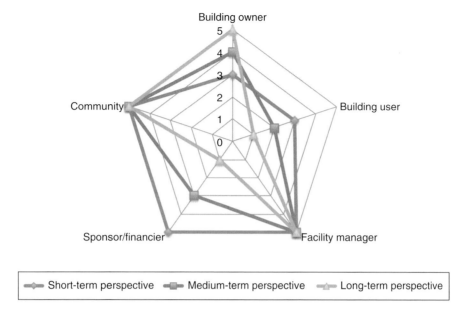

Figure 11.5 Example radar chart for stakeholder interest.

to add value to this asset. Further investigation is worthy and recommended. An 'X' marks the position of the property on the 2D format. The scores for each attribute can be presented using 'radar' charts for ease of understanding (an example based on stakeholder interest is provided in Figure 11.5).

11.4.3 After Intervention

Information was not readily available to recreate the performance of this property over its 124-year life. If it was, it is likely that it would have started near the upper retain/extend corner of the cube and moved slowly towards the lower reconstruct/dispose corner over time. This trend would have been reversed occasionally when refurbishment activities were undertaken (e.g. 1886, 1912, 1986 and 2008). In recent years prior to 2008, the property evidenced reasonable levels of physical condition but became surplus to its original industrial objectives, and its utilisation correspondingly decreased. This combination would have prompted adaptive reuse as a wise response to take advantage of the embedded physical life still remaining in the property. Eventual conversion to boutique office space enabled utilisation to be revitalised and the property to be given a new lease of life. Figure 11.6 shows the *iconCUR* model evaluation after intervention.

This case study highlights that adaptive reuse was the trending decision before the latest intervention, but 'recycle' was identified as a possible interim action. Lack of significant action would have resulted in a decrease in condition that would strengthen the decision to start recycling parts of the building for other uses (e.g. on other sites). Following the intervention, utilisation was the main performance change as the building is now fully occupied and

iconCUR
Property management matrix
Condition : Utilisation : Reward

iconCUR Assessment worksheet
88 George Street, Sydney (Australia)
Original construction: 1886   126 years old   (after proposed intervention)

High   < Utilisation   Low

	Renovate / preserve	Repair		Retain / extend
			Invest capital	
	Retrofit	Reconfigure		Refresh
			Natural decay	
	Reconstruct / dispose	Recycle		Reuse / adapt

Low   Condition >   High

Condition   3.90

	Weighting	Design standard 50%	Maintained service level 25%	Regulatory compliance 25%	(high) 100%
Structure	20%	4	4	4	4.00
Exterior envelope	30%	4	4	4	4.00
Interior finishes/fitout	20%	4	3	4	3.75
Engineering services	20%	4	3	4	3.75
External works	10%	4	3	5	4.00
	100%	4.00	3.50	4.10	3.90

Utilisation   2.88

	Weighting	Demand or relevance 40%	Fitness for purpose 40%	User satisfaction 20%	(moderate) 100%
Internal space (feca)	30%	4	3	3	3.40
External space (uca)	10%	0	2	0	0.80
Outdoor site area	10%	3	3	2	2.80
Equipment and fitout	30%	3	3	3	3.00
Engineering systems	20%	3	3	3	3.00
	100%	3.00	2.90	2.60	2.88

Collective utility   4.20

	Weighting	Economic performance 20%	Culture and heritage 40%	Environmental values 40%	(very high) 100%
Operational viability	30%	4	4	5	4.40
Locational context	20%	5	5	1	3.40
Risk and opportunity	20%	3	5	4	4.20
Asset valuation	30%	4	5	4	4.40
Profile/mission	10%	5	5	5	4.80
	100%	4.00	4.70	3.80	4.20

Stakeholder interest   3.80

	Weighting	Short-term perspective 50%	Medium-term perspective 25%	Long-term perspective 25%	(high) 100%
Building owner	40%	3	4	5	4.20
Building user	20%	3	2	1	1.80
Facility manager	20%	5	5	5	5.00
Sponsor/financier	10%	5	3	1	2.60
Community	20%	5	5	5	5.00
	100%	3.80	3.80	3.80	3.80

Reward   3.19

	Collective utility	Utilisation (y axis)	Stakeholder interest	Reward (z axis)	(high)
	4.20	X	3.80	3.80	

Asset performance   3.32

	Condition (x axis)	Utilisation (y axis)	Reward (z axis)	(good)
Property status	3.90	2.88	3.19	
Trending decision	5.00	5.00	5.00	Reconfigure / Retain/extend
Decision exposure				High risk
Proximity to corner		66%	65%	Far
Extra reward likelihood				Low

iconCUR that change is not recommended at the present time

Figure 11.6   *iconCUR model output (after intervention).*

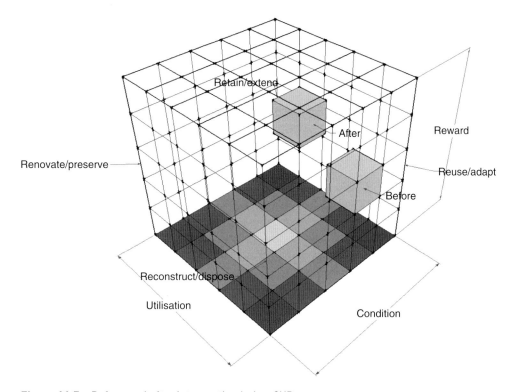

Figure 11.7 Before and after intervention in *iconCUR*.

an asset likely to be retained. Its reward value also increased (with the benefit of hindsight) from 2.28 to 3.19, or about 40% of its previous value. The condition value rose to 3.90 (10% more) and the utilisation value rose to 2.88 (200% more). Figure 11.7 shows that the decision to proceed with a 'green' adaptive reuse refurbishment improved all three criteria scores and thus elevated the overall project status within the *iconCUR* model.

11.5   Discussion

The *iconCUR* model is useful only if a property's coordinates can be quantified. Langston (2013) employed a weighted matrix approach to add detail so that measurements can be objective and justified. For each criterion or sub-criterion in the model, key attributes (A1–A3) were provided in columns and weighted according to their relevance (total must be 100%). Under each attribute, key elements (E1–E5) were provided in rows and weighted according to their influence (total also must be 100%). Performance was scored at the intersection of each attribute and element using a scale of 0–5 (0 = weak/nil and 5 = strong) in the context of relevant asset class expectations and accepted benchmarks, based on expert opinion.

Table 11.1 shows the impact of changes in weights to the calculation of coordinates for 88 George Street before intervention. Ten sets of weights are

Table 11.1   The effect of different weights (before intervention).

	A1(%)	A2(%)	A3(%)	E1(%)	E2(%)	E3(%)	E4(%)	E5(%)	Coordinate
Condition									
Original	50.00	25.00	25.00	20.00	30.00	20.00	20.00	10.00	**3.55**
A1 × 2	**100.00**	0.00	0.00	20.00	30.00	20.00	20.00	10.00	3.40
A2 × 2	37.50	**50.00**	12.50	20.00	30.00	20.00	20.00	10.00	3.51
A3 × 2	37.50	12.50	**50.00**	20.00	30.00	20.00	20.00	10.00	3.66
E1 × 2	50.00	25.00	25.00	**40.00**	22.50	15.00	15.00	7.50	3.66
E2 × 2	50.00	25.00	25.00	11.43	**60.00**	11.43	11.43	5.71	**3.74**
E3 × 2	50.00	25.00	25.00	15.00	22.50	**40.00**	15.00	7.50	3.48
E4 × 2	50.00	25.00	25.00	15.00	22.50	15.00	**40.00**	7.50	**3.29**
E5 × 2	50.00	25.00	25.00	17.78	26.66	17.78	17.78	**20.00**	3.60
All equal	**33.33**	**33.33**	**33.33**	20.00	**20.00**	20.00	20.00	**20.00**	3.60
								Mean	3.55
								CoV	3.83%
Utilisation									
Original	40.00	40.00	20.00	30.00	10.00	10.00	30.00	20.00	**0.96**
A1 × 2	**80.00**	13.33	6.67	30.00	10.00	10.00	30.00	20.00	0.85
A2 × 2	13.33	**80.00**	6.67	30.00	10.00	10.00	30.00	20.00	1.19
A3 × 2	30.00	30.00	**40.00**	30.00	10.00	10.00	30.00	20.00	0.87
E1 × 2	40.00	40.00	20.00	**60.00**	5.71	5.71	17.15	11.43	**1.23**
E2 × 2	40.00	40.00	20.00	26.67	**20.00**	8.88	26.67	17.78	0.94
E3 × 2	40.00	40.00	20.00	26.67	8.88	**20.00**	26.67	17.78	0.85
E4 × 2	40.00	40.00	20.00	17.15	5.71	5.71	**60.00**	11.43	0.89
E5 × 2	40.00	40.00	20.00	22.50	7.50	7.50	22.50	**40.00**	0.92
All equal	**33.33**	**33.33**	**33.33**	**20.00**	**20.00**	**20.00**	**20.00**	20.00	**0.73**
								Mean	0.94
								CoV	16.29%
Collective utility									
Original	20.00	40.00	40.00	30.00	20.00	20.00	20.00	10.00	**3.00**
A1 × 2	**40.00**	30.00	30.00	30.00	20.00	20.00	20.00	10.00	3.00
A2 × 2	13.33	**80.00**	6.67	30.00	20.00	20.00	20.00	10.00	**3.66**
A3 × 2	6.67	13.33	**80.00**	30.00	20.00	20.00	20.00	10.00	**2.40**
E1 × 2	20.00	40.00	40.00	**60.00**	11.43	11.43	11.43	5.71	2.74
E2 × 2	20.00	40.00	40.00	22.50	**40.00**	15.00	15.00	7.50	3.10
E3 × 2	20.00	40.00	40.00	22.50	15.00	**40.00**	15.00	7.50	3.05
E4 × 2	20.00	40.00	40.00	22.50	15.00	15.00	**40.00**	7.50	3.00
E5 × 2	20.00	40.00	40.00	26.66	17.78	17.78	17.78	**20.00**	3.07
All equal	**33.33**	**33.33**	**33.33**	**20.00**	20.00	20.00	20.00	**20.00**	3.13
								Mean	3.02
								CoV	10.45%
Stakeholder interest									
Original	50.00	25.00	25.00	40.00	20.00	10.00	10.00	20.00	**3.80**
A1 × 2	**100.00**	0.00	0.00	40.00	20.00	10.00	10.00	20.00	3.80
A2 × 2	37.50	**50.00**	12.50	40.00	20.00	10.00	10.00	20.00	3.80
A3 × 2	37.50	12.50	**50.00**	40.00	20.00	10.00	10.00	20.00	3.80
E1 × 2	50.00	25.00	25.00	**80.00**	6.67	3.33	3.33	6.67	3.77
E2 × 2	50.00	25.00	25.00	30.00	**40.00**	7.50	7.50	15.00	**3.41**

Table 11.1   (*Cont'd*).

	A1(%)	A2(%)	A3(%)		E1(%)	E2(%)	E3(%)	E4(%)	E5(%)	Coordinate
E3 × 2	50.00	25.00	25.00		35.56	17.78	**20.00**	8.88	17.78	3.93
E4 × 2	50.00	25.00	25.00		35.56	17.78	8.88	**20.00**	17.78	3.77
E5 × 2	50.00	25.00	25.00		30.00	15.00	7.50	7.50	**40.00**	**4.10**
All equal	**33.33**	**33.33**	**33.33**		**20.00**	20.00	**20.00**	**20.00**	20.00	3.80
									Mean	3.80
									CoV	4.48%
Reward										
Original										**2.28**
A1 × 2										2.28
A2 × 2										**2.78**
A3 × 2				Collective utility x stakeholder interest						**1.82**
E1 × 2				———————————————————						2.07
E2 × 2				5						2.12
E3 × 2										2.40
E4 × 2										2.26
E5 × 2										2.51
All equal										2.38
									Mean	2.29
									CoV	11.38%

tested. The first set reflects the original weights (i.e. assumed decision-maker preferences). The next three sets double the importance of attributes A1–A3, respectively, while the remaining weights are reduced in proportion to their original values so that the total continues to equal 100%. The next five sets double the importance of elements E1–E5, respectively, while the remaining weights are reduced in proportion to their original values so the total similarly continues to equal 100%. The final set assumes that both attributes and elements are evenly weighted. Table 11.2 shows the impact of changes in weights to the calculation of coordinates after intervention. The percentage weights in bold highlight the changes made in each set, while the spatial coordinate in the last column is the product of *score* × *weight*. The original coordinate and the highest and lowest coordinates resulting from the test are also shown in bold. These values are carried forward for further analysis of decision impact.

The analysis indicates that the mean coordinate is close to the original coordinate under all scenarios, and the ten results computed for each criterion or sub-criterion have very low coefficients of variation (CoV), showing little dispersion around the mean. The coordinates can be used to plot the asset's overall performance (i.e. status) at any point in time in 3D space. The original coordinates suggested that adaptive reuse was the appropriate intervention strategy, and an improvement in all three variables occurred post-intervention.

Table 11.2 The effect of different weights (after intervention).

	A1(%)	A2(%)	A3(%)	E1(%)	E2(%)	E3(%)	E4(%)	E5(%)	Coordinate
Condition									
Original	50.00	25.00	25.00	20.00	30.00	20.00	20.00	10.00	**3.90**
A1 × 2	**100.00**	0.00	0.00	20.00	30.00	20.00	20.00	10.00	**4.00**
A2 × 2	37.50	**50.00**	12.50	20.00	30.00	20.00	20.00	10.00	**3.76**
A3 × 2	37.50	12.50	**50.00**	20.00	30.00	20.00	20.00	10.00	3.99
E1 × 2	50.00	25.00	25.00	**40.00**	22.50	15.00	15.00	7.50	3.93
E2 × 2	50.00	25.00	25.00	11.43	**60.00**	11.43	11.43	5.71	3.94
E3 × 2	50.00	25.00	25.00	15.00	22.50	**40.00**	15.00	7.50	3.86
E4 × 2	50.00	25.00	25.00	15.00	22.50	15.00	**40.00**	7.50	3.86
E5 × 2	50.00	25.00	25.00	17.78	26.66	17.78	17.78	**20.00**	3.91
All equal	**33.33**	**33.33**	**33.33**	20.00	**20.00**	20.00	20.00	**20.00**	3.87
								Mean	3.90
								CoV	1.77%
Utilisation									
Original	40.00	40.00	20.00	30.00	10.00	10.00	30.00	20.00	**2.88**
A1 × 2	**80.00**	13.33	6.67	30.00	10.00	10.00	30.00	20.00	2.96
A2 × 2	13.33	**80.00**	6.67	30.00	10.00	10.00	30.00	20.00	2.89
A3 × 2	30.00	30.00	**40.00**	30.00	10.00	10.00	30.00	20.00	2.81
E1 × 2	40.00	40.00	20.00	**60.00**	5.71	5.71	17.15	11.43	**3.10**
E2 × 2	40.00	40.00	20.00	26.67	**20.00**	8.88	26.67	17.78	2.65
E3 × 2	40.00	40.00	20.00	26.67	8.88	**20.00**	26.67	17.78	2.87
E4 × 2	40.00	40.00	20.00	17.15	5.71	5.71	**60.00**	11.43	2.93
E5 × 2	40.00	40.00	20.00	22.50	7.50	7.50	22.50	**40.00**	2.91
All equal	**33.33**	**33.33**	**33.33**	**20.00**	**20.00**	**20.00**	**20.00**	20.00	**2.53**
								Mean	2.85
								CoV	5.62%
Collective utility									
Original	20.00	40.00	40.00	30.00	20.00	20.00	20.00	10.00	**4.20**
A1 × 2	**40.00**	30.00	30.00	30.00	20.00	20.00	20.00	10.00	4.15
A2 × 2	13.33	**80.00**	6.67	30.00	20.00	20.00	20.00	10.00	**4.55**
A3 × 2	6.67	13.33	**80.00**	30.00	20.00	20.00	20.00	10.00	**3.93**
E1 × 2	20.00	40.00	40.00	**60.00**	11.43	11.43	11.43	5.71	4.29
E2 × 2	20.00	40.00	40.00	22.50	**40.00**	15.00	15.00	7.50	4.00
E3 × 2	20.00	40.00	40.00	22.50	15.00	**40.00**	15.00	7.50	4.20
E4 × 2	20.00	40.00	40.00	22.50	15.00	15.00	**40.00**	7.50	4.25
E5 × 2	20.00	40.00	40.00	26.66	17.78	17.78	17.78	**20.00**	4.27
All equal	**33.33**	**33.33**	**33.33**	**20.00**	20.00	20.00	20.00	**20.00**	4.20
								Mean	4.20
								CoV	3.95%
Stakeholder interest									
Original	50.00	25.00	25.00	40.00	20.00	10.00	10.00	20.00	**3.80**
A1 × 2	**100.00**	0.00	0.00	40.00	20.00	10.00	10.00	20.00	3.80
A2 × 2	37.50	**50.00**	12.50	40.00	20.00	10.00	10.00	20.00	3.80
A3 × 2	37.50	12.50	**50.00**	40.00	20.00	10.00	10.00	20.00	3.80
E1 × 2	50.00	25.00	25.00	**80.00**	6.67	3.33	3.33	6.67	3.77
E2 × 2	50.00	25.00	25.00	30.00	**40.00**	7.50	7.50	15.00	**3.41**

Table 11.2   (*Cont'd*).

	A1(%)	A2(%)	A3(%)	E1(%)	E2(%)	E3(%)	E4(%)	E5(%)	Coordinate
E3 × 2	50.00	25.00	25.00	35.56	17.78	**20.00**	8.88	17.78	3.93
E4 × 2	50.00	25.00	25.00	35.56	17.78	8.88	**20.00**	17.78	3.77
E5 × 2	50.00	25.00	25.00	30.00	15.00	7.50	7.50	**40.00**	**4.10**
All equal	**33.33**	**33.33**	**33.33**	**20.00**	20.00	**20.00**	**20.00**	20.00	3.80
								Mean	3.80
								CoV	4.48%
Reward									
Original									**3.19**
A1 × 2									3.15
A2 × 2									3.46
A3 × 2			Collective utility x stakeholder interest						2.99
E1 × 2			5						3.23
E2 × 2									**2.73**
E3 × 2									3.30
E4 × 2									3.20
E5 × 2									**3.50**
All equal									3.19
								Mean	3.19
								CoV	6.87%

Taking the base estimate (original), lowest (worst case) and highest (best case) spatial coordinates as characteristic of the resulting range, it is discovered that these three sets reflect the mean, lower and upper values, respectively, out of the 81 possible combinations of coordinates both before and after intervention. The computed asset performance score (out of 5) before intervention ranges from 1.89 to 2.66, with a mean equal to the original value of 2.26. The CoV is 7.52% ($n = 81$). The computed asset performance after intervention ranges from 2.99 to 3.61, with a mean of 3.30 (i.e. very close to the original value of 3.32). The CoV in this case is 4.04% ($n = 81$).

The assessment of decision risk in *iconCUR* can be achieved by computing the distance between a property's coordinates and the various matrix 'hotspots'. The risk attached to an alternative (or outcome) is given by d_1 (the distance from the centre of the cube to the property coordinates). The smaller the distance, the higher the risk (measured in 2D space). Low risk is defined as $d_1 \geq 2.5$ (i.e. decision-maker should proceed), moderate risk is defined as $1.5 \leq d_1 < 2.5$ (i.e. decision-maker should consider interim actions such as retrofit, recycle, refresh and repair), and high risk is defined as $d_1 < 1.5$ (i.e. decision-maker should make only minor reconfiguration changes or wait).

Similarly, the priority of a property towards a particular outcome is defined by d_2 (the distance from an upper cube corner to the property coordinates), where the smaller the distance, the higher the priority (measured in 3D space). This can be described as very high ($d_2 \leq 1$), high ($1 < d_2 \leq 2$),

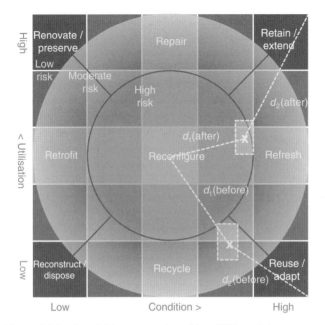

Figure 11.8   Spatial interpretation of *iconCUR* model.

moderate ($2 < d_2 \leq 3$), low ($3 < d_2 \leq 4$) and very low ($d_2 > 4$) priority. This needs to be considered in the context of potential value add.

Figure 11.8 summarises this approach for 88 George Street and shows the calculation of d_1 and d_2 both before and after intervention, as well as the zones of possible variation. It should be noted that the variation zones are quite small.

Interestingly, the property status moved from moderate risk (trending towards the reuse/adapt corner) to high risk (trending towards the retain/ extend corner). In the latter case, the coordinates are very close to the moderate risk zone, suggesting that a higher condition standard would have made it so.

From the three selected combinations of spatial coordinates, values for d_1 before intervention are all shown to be moderate risk, and out of the 81 possible combinations, all reflect moderate risk except for 9 that are high risk, with d_1 in the range 1.50–2.16. Values for d_1 after intervention comprise high risk (original and worst case) and moderate risk (best case), and out of the 81 possible combinations, 45 reflect high risk and 36 reflect moderate risk, with d_1 in the range 1.26–1.62. In every one of the 81 possible combinations of spatial coordinates before intervention, the trending decision is shown as 'reuse/adapt', and in every one of the 81 possible combinations of spatial coordinates after intervention, the trending decisions is 'retain/ extend'. This indicates that the robustness of the model is not unduly affected by the choice of criterion weights, although the level of risk can change between worst-case, original and best-case estimates. In other words, the criterion weights had no effect on the intervention strategy selected.

Values for d_2 before intervention for the three selected combinations of spatial coordinates are shown to be low priority (original and worst case) and moderate priority (best case), with d_2 in the range 2.47–3.97. After intervention the priority is still low (worst case) and moderate (best case), but the priority (original) has increased from low to moderate, with d_2 in the range 2.49–3.61. Despite the improvement in asset performance scores somewhere between 36% and 59%, what this in fact means is that the property is unlikely to make a compelling case for retain/extend when compared with other office building options that are newer and more purposefully designed for this function. Nevertheless, retain/extend remains the current trending decision for this property.

While this chapter investigates merely one case study before and after intervention, the results nevertheless suggest that routine analysis of the impact of criterion weights may be needed only in unusual circumstances where the normal influence of central tendency may be absent (i.e. if deliberate bias in preference setting is suspected). These results at least provide some counterbalance to the assertion that criterion weighting in MCDA can lead to erroneous decisions.

11.6   Conclusion

The *iconCUR* model is a practical demonstration of the ability to map decision criteria (condition, utilisation, reward) and sub-criteria (collective utility, stakeholder interest) with attributes (e.g. design standard, maintained service level, regulatory compliance) and alternatives (retain/extend, renovate/preserve, reuse/adapt, reconstruct/dispose) using a classic MCDA hierarchical approach of weighted scores. Change is recommended when the trending decision (as determined by matrix hotspots) and property status (as determined by current spatial coordinates) coalesce, given that reward value add is anticipated and judged to be worth the effort. The ability to track decisions over time also enables the veracity of those decisions to be ultimately confirmed by history. The deployment of this type of approach, at least, documents the decision-making process involved in important property management strategies and exposes the frequently implicit preferences of decision-makers. These preferences are shown not to be as critical as might have been suspected. This adds confidence to the use of MCDA in facilities management decisions.

Notes

1   See http://visualpromethee.com/pptVisual.html for a useful summary of modern 'visual MCDA'.

2   Note that just a few years later, Australia's first 6-star *Green Star* heritage-listed refurbishment was carried out at 39 Hunter Street, Sydney, although technically it was not an adaptive reuse conversion. Legion House at 161 Castlereagh Street, Sydney, is currently undergoing adaptive reuse and is touted to be the greenest heritage-listed conversion in the world upon completion (http://www.legion house.com.au/index.html).

References

Aguilar, S.C. (2009) Decision making and brownfield development, *E + M Ekonomie a Management*, 2(3), 19–32.

Alanne, K. (2004) Selection of renovation actions using multi-criteria 'knapsack' model, *Automation in Construction*, 13(3), 377–391.

Aznar, J., Ferrís-Oñate, J. and Guijarro, F. (2010) An ANP framework for property pricing combining quantitative and qualitative attributes, *Journal of the Operational Research Society*, 61(5), 740–755.

Belton, V. and Stewart, T.J. (2002) *Multiple criteria decision analysis: an integrated approach*, Boston, MA: Kluwer Academic Publishing.

Berbel, J. and Zamora, R. (1995) An application of MOP and GP to wildlife management (deer), *Journal of Environmental Management*, 44(1), 29–38.

Bostenaru Dan, M.D. (2004) Multi-criteria decision model for retrofitting existing buildings, *Natural Hazards and Earth System Sciences*, 4, 485–499.

Cebi, S. and Kahraman, C. (2010) Fuzzy multicriteria group decision making for real estate investments, *Journal of Systems and Control Engineering*, 224(14), 457–470.

Douglas, J. (2006) *Building adaptation* (2nd edition), London: Elsevier.

Ensslin, S.R., Ensslin, L. and de Oliveira Lacerda, RT (2011) A performance measurement framework in portfolio management, *Management Decision*, 49(4), 648–668.

Gann, D.M., Salter, A.J. and Whyte, J.K. (2003) Design quality indicator as a tool for thinking, *Building Research and Information*, 31(5), 318–333.

Georgiou, I. (2008) Making decisions in absence of clear facts, *European Journal of Operational Research*, 185(1), 299–321.

Henig, M.I. and Buchanan, J.T. (1996) Solving MCDM problems: process concepts, *Journal of Multi-criteria Decision Analysis*, 5, 3–21.

Herath, G. (2004) Incorporating community objectives in improved wetland management: the use of analytic hierarchy process, *Journal of Environmental Management*, 70(3), 263–273.

Kaklauskas, A., Zavadskas, E.K. and Raslanas, S. (2005) Multivariant design and multiple criteria analysis of building refurbishments, *Energy and Buildings*, 37(4), 361–372.

Kincaid, D. (2002) *Adapting buildings for changing uses: guidelines for change of use refurbishment*, London: Spon Press.

Langston, C. (2005) *Life-cost approach to building evaluation*, Amsterdam: Elsevier.

Langston, C. (2011) Green adaptive reuse: issues and strategies for the built environment, in *Modeling risk management in sustainable construction*, Wu, D.D. ed., Berlin: Springer, 199–210.

Langston, C. (2012a) The role of coordinate-based decision-making in the evaluation of sustainable built environments, *Construction Management and Economics*, 31(1), 62–77.

Langston, C. (2012b) The impact of criterion weights in facilities management decision-making: an Australian case study, *Facilities*, 31(7/8), 270–289.

Langston, C. and Smith, J. (2012) Modelling property management decisions using iconCUR, *Automation in Construction*, 22, 406–413.

Laukkanen, S., Palander, T. and Kangas, J. (2004) Applying voting theory in participatory decision support for sustainable timber harvesting, *Canadian Journal of Forest Research*, 34(7), 1511–1524.

Mardle, S.J. and Pascoe, S. (2002) Modelling the effects of trade-offs between long and short-term objectives in fisheries management, *Journal of Environmental Management*, 65(1), 49–62.

Mareschal, B. and Brans, J.P. (1988) Geometrical representations for MCDA: the GAIA module, *European Journal of Operational Research*, 34, 69–77.

Martin, W.E., Shields, D.J., Tolwinski, B. and Kent, B. (1996) An application of social choice theory to U.S.D.A. forest service decision making, *Journal of Policy Modeling*, 18(6), 603–621.

Mergias, I., Moustakas, K., Papadopoulos, A. and Loizidou, M. (2007) Multi-criteria decision aid approach for the selection of the best compromise management scheme for ELVs: the case of Cyprus, *Journal of Hazardous Materials*, 147(3), 706–717.

Ribeiro, F.L. and Videira, S.I. (2008) Management of the built heritage in the Lisbon's central downtown, *International Journal of Housing Markets and Analysis*, 1(2), 110–124.

Roper, K.O., Kim, J.H., Juan, Y.K. and Castro-Lacouture, D. (2010) Optimal decision making on urban renewal projects, *Management Decision*, 48(2), 207–224.

Rosato, P., Giove, S. and Breil, M. (2010) An application of multicriteria decision making to built heritage: the redevelopment of Venice Arsenale, *Journal of Multi-Criteria Decision Analysis*, 17(3–4), 85–99.

Rosenfeld, Y. and Shohet, I.M. (1999) Decision support model for semi-automated selection of renovation alternatives, *Automation in Construction*, 8(4), 503–510.

Stewart, T.J., Janssen, R. and Van Herwijnen, M. (2004) A genetic algorithm approach to multi-objective land use planning, *Computers and Operations Research*, 31(14), 2293–2313.

Subbu, R., Russo, G., Chalermkraivuth, K. and Celaya, J. (2007) Multi-criteria set partitioning for portfolio management: a visual interactive model, in proceedings of Computational Intelligence in Multicriteria Decision Making Conference, Honolulu, HI, April 1–5, pp. 166–171.

Sun, Y., Fidge, C. and Ma, L. (2008) A generic split process model for asset management decision-making. Available at http://sky.fit.qut.edu.au/~fidgec/Publications/sun08a.pdf. Accessed on 8 August 2013

Trinkhaus, H.L. and Hanne, T. (2003) KnowCube for MCDM: visual and interactive support for multicriteria decision making, *Computers and Operations Research*, 32(5), 1289–1309.

Tupenaite, L., Mickaityte, A., Zavadskas, E. and Kaklauskas, A. (2008) The concept model of sustainable buildings refurbishment, *International Journal of Strategic Property Management*, 12(1), 53–68.

Urbanaviciene, V., Kaklauskas, A., Zavadskas, E. and Seniut, M. (2009) The web-based real estate multiple criteria negotiation decision support system: a new generation of decision support systems, *International Journal of Strategic Property Management*, 13(3), 267–286.

Vadrevu, K.P., Eaturu, A. and Badarinath, K.V.S. (2010) Fire risk evaluation using multicriteria analysis: a case study, *Environmental Monitoring and Assessment*, 166(1), 223–239.

Yau, Y. (2009) Multi-criteria decision making for urban built heritage conservation: application of the analytic hierarchy process, *Journal of Building Appraisal*, 4(3), 191–205.

12

Designing for Future Adaptive Reuse

12.1 Introduction

Existing buildings that are either obsolete or rapidly approaching disuse and potential demolition are a 'mine' of raw materials for new projects, a concept described by Chusid (1993) as 'urban ore'. Rather than extracting these raw materials during demolition or deconstruction and assigning them to new applications, it is more effective to leave the basic structure and fabric of the building intact and adaptively change its use. Breathing 'new life' into existing buildings carries with it numerous environmental and social benefits and helps to retain our cultural heritage. To date, a focus on economic factors alone has contributed to destruction of some buildings well short of their notional physical lives.

Planned adaptive reuse is advanced as an emerging and fundamental design consideration for all new projects in the context of national climate change and emission reduction strategies. The reuse of obsolete buildings without extensive demolition provides significant opportunity for the conservation of resources and the associated energy embedded in material manufacture and assembly (see Figure 12.1). However, one critical area of investigation remains. An evaluation tool is needed that helps guide proposed design so it can be optimised for future adaptive reuse from the outset. In this way new construction can positively contribute to long-term resource efficiency in a wider sense than merely recurrent operational performance might suggest.

Sustainable Building Adaptation: Innovations in Decision-Making, First Edition.
Sara J. Wilkinson, Hilde Remøy and Craig Langston.
© 2014 John Wiley & Sons, Ltd. Published 2014 by John Wiley & Sons, Ltd.

Figure 12.1   Town Hall (1880), UNESCO World Cultural Heritage City, George Town, Malaysia.

12.2   Rationale

Refurbishment can of itself take many forms, ranging from simple cosmetic decoration to significant reconstruction and renewal. Sometimes the buildings are in quite good condition, but the services and technology within them are outdated, in which case retrofit may be appropriate. If a particular function is no longer relevant or desired, buildings may be converted to a new purpose altogether.

Older buildings often have a character that can contribute to the ongoing culture of a society and conserve aspects of its history. The preservation of these buildings is important and maintains their intrinsic heritage and cultural values. Facilities managers are frequently faced with decisions about whether to rent or buy, whether to extend or sell and whether to refurbish, demolish or construct. Usually these decisions are based on financial matters, but there are other issues that should bear on the final choice, including environmental and social impacts.

For a wide range of reasons, buildings can become obsolete long before their physical life has come to an end. Investing in long-lived buildings may be suboptimal if their useful life falls well short of their physical life. It is wise to design future buildings for change by making them more flexible yet with sufficient structural integrity to support alternative functional use. The development of a design-rating scheme for adaptation potential will enable building designers to understand the long-term impacts of their decisions prior to construction and thus enable optimisation for adaptive reuse to occur from the outset. As adaptive reuse potential (ARP) already embodies financial, social and environmental criteria, the rating scheme will extend traditional operational considerations such as energy performance to include churn, retrofit, refurbishment and renewal considerations.

Atkinson (1988) modelled the process of obsolescence and renewal (of housing stock) and developed a 'sinking stack' theory to explain the phenomenon. Comparing total building stock over time produces a rising profile in total stock (accumulating via new construction each year) stratified according to building age (older buildings are at lower layers in the profile strata). New stock is added annually to the top of the stack. It degenerates over time and gradually sinks towards the bottom of the stack as new buildings are created and older ones demolished. If little new construction is added, then the entire building stock will age, and greater resources will be required to maintain overall quality and amenity levels. Certain layers in the stack are likely to represent periods of poor-quality construction, and these layers age more rapidly and absorb greater maintenance resources (Ness and Atkinson 2001). Each layer in the stack reduces in height with the passage of time. Only the top layer grows because it represents the current rate of new construction. The net effect is a sinking of the stack, a phenomenon that occurs whether or not sufficient maintenance takes place.

From an environmental perspective, it is preferable to minimise new additions to the stack but at the same time remove those layers of poorer quality stock that absorb excessive maintenance and operating resources. Increased resources should be allocated to maintenance of those better-quality layers of the stack. Atkinson developed computer models that illustrated the sinking effect dynamically for given input parameters. The philosophy of 'minimum decay' (Atkinson 1988) involves retarding the rate of obsolescence and replacement, or in other words, slowing down the sinking of the stack by decreasing the consumption of new resources, and assigning increased resources to maintenance and refurbishment activities. Where this can be linked to improving operating energy efficiency and comfort, the saving in embodied energy (i.e. energy already invested in manufacture and construction processes) is substantial. The development of a design-rating scheme for adaptation potential enables future additions to the stack to be of higher resource value through a planned renewal strategy and a focus on 'long life, loose fit and low energy' that is the philosophical underpinning for sustainable built environment performance.

Adaptive reuse is a special form of building refurbishment that poses quite difficult challenges for designers. Significantly changing the class (or functional classification) of a building during its life cycle introduces new regulatory conditions and perhaps requires zoning consent. But there are clear economic, environmental and social benefits that can make this option attractive to developers, such as increases in floor space ratios and concessions received for pursuing government policy directions by regenerating derelict public assets. In recent times, redundant city office buildings have been converted into high-quality residential apartments, bringing people back to cities and in the process revitalising urban precincts.

Adaptive reuse, as a process, has been applied to many types of facilities, including defence estates (e.g. Doak 1999; Van Driesche and Lane 2002), airfields (e.g. Gallent et al. 2000), government buildings (e.g. Abbotts et al. 2003) and industrial buildings (e.g. Ball 1999). Internationally, adaptive reuse of historic buildings is seen as fundamental to sound government policy and sustainable development – for example, in Atlanta, USA (Newman 2001), Canada (Brandt 2006), Hong Kong (Poon 2001), North Africa (Leone 2003) and Australia (McLaren 1996; Maggs 1999).

Adaptive reuse can be quite dramatic. For example, conversion of disused industrial factories into shopping centres or churches into restaurants can redefine and revitalise districts. Facilities managers should be conscious of adaptive reuse solutions to redundant space and continually think about more productive uses for premises that are underutilised. It is therefore critical that mechanisms are in place to ensure that new buildings represent value to society (their long-term stakeholder) rather than their short-term custodians or brokers. Unmasking the social 'costs' of renewal can provide strong incentives for a transition to more sustainable energy use, less profligate use of new materials, reduced waste and greater service from constructed building stock.

Therefore, the significance of this research lies in the empowerment of building designers to understand the long-term impacts of their decisions and so be better placed to strive for solutions that contribute to ecological sustainability and help mitigate against further climate change pressures. The information provided by this research facilitates the design and justification of buildings suitable for adaptive reuse before costly or irreversible decisions are made. The balance between project feasibility, environmental impact and social benefit can thus be objectively evaluated in the light of project-specific constraints and stakeholder interests. Projects that exhibit best practice in ARP can be readily identified via a user-friendly star-rating system.

The innovation of this research lies with the reverse engineering of the ARP model (Langston 2008) discussed in Chapter 9, so that design pathways can be readily evaluated to optimise building proposals and be more aligned to long-term societal goals. In the future, designers will be able to receive guidance on the effectiveness of their proposals towards achievement of true resource efficiency, explicitly taking account of embodied energy of construction, churn, retrofit, refurbishment and renewal over the entire life cycle, and benchmark this against best practice. It is clever because it adopts a well-known star-rating paradigm yet extends it beyond traditional operating performance issues to consider multiple 'lives' for the building that potentially encourage new asset stock to be designed for longevity – an obvious objective that modern society seems to have largely forgotten.

Currently, theory and practice are aligned in the sense that there are consistent gaps with regard to the implementation of sustainable developments. Efforts to align project decision-making during the design phase with environmental performance knowledge are at best based on incomplete data (Lenzen and Treloar 2003). This research addresses the problem by collection and analysis of data to explain the link between design strategies and adaptive reuse success and make this explicit for the first time.

Two research outcomes are clear. Improvement in adaptation preparedness for new buildings leads to more efficient long-term use of domestic resources and less demand on our environment, as well as elevates the performance of buildings in lower strata of the built environment 'stack'. Furthermore, national heritage is conserved through reusing buildings of greater permanence that, although outliving their original purpose, are still capable of making significant contributions to our urban landscape via a built-in propensity for future adaptive reuse. The outcomes of this research will also help to change the focus for the property industry, as it must, from short-term solutions to longer-term refurbishment and adaptive reuse, and thus make more substantial contributions to national and international sustainability goals. Reduced environmental footprints through lower embodied and operational energy demand in built facilities, decreased greenhouse gas emissions, less waste to landfill and making better use of what we already have are critical to national and international sustainability targets and policy. The greenest buildings are the ones we already have (Jacobs 1961).

12.3   *AdaptSTAR* Framework

AdaptSTAR is an attempt to rate new building design for future 'adaptivity'. This rating is done normally when the project is in its design phase, although it can be applied in hindsight based on the latent conditions before a proposed intervention takes place.

The concept of the *adaptSTAR* design-rating scheme for adaptation potential (Conejos 2013) was founded on the categories of obsolescence discussed in Chapter 9. Each category was broken down into sub-criteria that were assembled from the literature (see Table 12.1) and from expert interviews with the design teams of 11 award-winning adaptive reuse conversions in New South Wales and the Australian Capital Territory (ACT), as well as a pilot study involving the Melbourne GPO in Victoria. The sub-criteria were then rated by a sample of practising Australian architects experienced in adaptive reuse work in order to determine the relative importance of each sub-criterion, which then led to the weight of each respective obsolescence category being computed. The results from this method are provided in Table 12.2. Note that the criterion weight is calculated from a five-point Likert scale (strongly disagree = 1, disagree = 2, neutral = 3, agree = 4, strongly agree = 5), while the consensus score is

Table 12.1   Summary of literature review.

Criterion	Previous research
Long life (physical)	
Structural integrity: structural design of the building to cater future uses and loads	Pevsner (1975), Osbourne (1985), Wittkower (1988), Grammenos and Russell (1997), Russell and Moffat (2001), Graham (2005), Davison et al. (2006), Douglas (2006), Siddiqi (2006), Gorse and Highfield (2009), Wilkinson et al. 2009, Horvath (2010), Yudelson (2010)
Material durability: durability of the building asset	Osbourne (1985), Grammenos and Russell (1997), Queensland Government (2000), Douglas (2006), UNEP (2007), Vakili-Ardebili (2007), Prowler (2008), Caroon (2010)
Workmanship: quality of craftsmanship of structure and finishes	Osbourne (1985), Whimster (2008)
Maintainability: building's capability to conserve operational resources	Osbourne (1985), City of New York (1999), Douglas (2006), Vakili-Ardebili (2007), Carter and Fortune (2008), Prowler (2008), Caroon (2010), Horvath (2010), Nakib (2010)
Design complexity: various geometries associated with the building's design and innovation	Grammenos and Russell (1997), Russell and Moffat (2001), Browne (2006)
Prevailing climate: changing climatic conditions	Wilson and Ward (2009)
Foundation: differential settlement and substrata movement	Osbourne (1985), UNEP (2007)
Location (economic)	
Population density: location within major city, CBD, etc.	Carter and Fortune (2008), Langston et al. (2008)
Market proximity: distance to major city, CBD, etc.	Campbell (1996), Fealy (2006), Prowler (2008), Wilkinson et al. (2009), Caroon (2010)
Transport infrastructure: availability and access	Heath (2001), Peiser and Schmitz (2007), UNEP (2007), Carter and Fortune (2008), Prowler (2008), Horvath (2010)
Site access: proximity or link to access roads, parking and communal facilities, etc.	Heath (2001), Peiser and Schmitz (2007), UNEP (2007), Carter and Fortune (2008), Prowler (2008), Wilkinson et al. (2009), Horvath (2010)
Exposure: views, privacy	Campbell (1996), Browne (2006), Fealy (2006)
Planning constraints: site selection, planning, neighbourhood and building design, etc.	City of New York (1999), Carter and Fortune (2008), Langston et al. (2008), Prowler (2008)
Plot size: built area, spatial proportions, enclosure, etc.	Campbell (1996), Heath (2001), Prowler (2008), Solomon (2008), Wilkinson et al. (2009)
Loose fit (functional)	
Flexibility: space capability to change according to newly required needs, plug-and-play elements, etc.	Grammenos and Russell (1997), Habraken (1998), City of New York (1999), Russell and Moffat (2001), Arge (2005), Graham (2005), Douglas (2006), UNEP (2007), Vakili-Ardebili (2007), Carter and Fortune (2008), Langston et al. (2008), Prowler (2008), Remøy et al. (2009a), Tobias and Vavaroutsos (2009), Wilkinson et al. 2009, Caroon (2010), Horvath (2010), Lehmann (2010), Nakib (2010), Zeiler et al. (2010)

(*continued*)

Table 12.1   (Cont'd).

Criterion	Previous research
Disassembly: options for reuse, recycle, demountable systems, deconstruction, modularity, etc.	City of New York (1999), Queensland Government (2000), Ness and Atkinson (2001), Russell and Moffat (2001), Graham (2005), Vakili-Ardebili (2007), Prowler (2008), Rabun and Kelso (2009), Tobias and Vavaroutsos (2009), Caroon (2010), Nakib (2010)
Spatial flow: mobility, open plan, fluid and continuous	Davison et al. (2006), Horvath (2010), Nakib (2010), Zeiler et al. (2010)
Convertibility: divisibility, elasticity, multi-functionality	City of New York (1999), Russell and Moffat (2001), Arge (2005), Nakib (2010)
Atria: open areas, interior gardens, etc.	Whimster (2008)
Structural grid: ideal and economical limit of span and fully interchangeable	Grammenos and Russell (1997), Russell and Moffat (2001), Arge (2005), Rabun and Kelso (2009), Remøy et al. (2009a)
Service ducts and corridors: vertical circulation, service elements, raised floors, etc.	Grammenos and Russell (1997), City of New York (1999), Russell and Moffat (2001), Davison et al. (2006), Prowler (2008), Rabun and Kelso (2009), Gilder (2010)
Low energy (technological)	
Orientation: microclimate siting, prevailing winds, sunlight	Park (1998), Douglas (2006), Shaw et al. (2007), UNEP (2007), Carter and Fortune (2008), Dittmark (2008), Prowler (2008), Knaack and Klein (2009), GBCA (2010), Appleby (2011)
Glazing: sunlight glare control and regulate internal temperatures, etc.	City of New York (1999), Douglas (2006), GBCA (2010), Appleby (2011)
Insulation and shading: thermal mass, sunshades, automated blinds, etc.	Osbourne (1985), Douglas (2006), Levine et al. (2007), UNEP (2007), Carter and Fortune (2008), Holborrow (2008), Prowler (2008), Knaack and Klein (2009), Tobias and Vavaroutsos (2009), Wilkinson et al. 2009, Farrell (2010), GBCA (2010), Lehmann (2010), Appleby (2011)
Natural lighting: inclusion for natural daylight, efficient lighting systems, etc.	Osbourne (1985), Park (1998), City of New York (1999), Queensland Government (2000), Davison et al. (2006), Douglas (2006), Levine et al. (2007), Shaw et al. (2007), Holborrow (2008), Tobias and Vavaroutsos (2009), Caroon (2010), GBCA (2010), Appleby (2011)
Natural ventilation: optimise airflow, quality fresh air, increase ambient air intake, etc.	Osbourne (1985), Park (1998), City of New York (1999), Queensland Government (2000), Ness and Atkinson (2001), Douglas (2006), Shaw et al. (2007), Holborrow (2008), Prowler (2008), Tobias and Vavaroutsos (2009), Wilson and Ward (2009), Caroon (2010), GBCA (2010), Horvath (2010), Appleby (2011)
Building management systems: monitor and control building operations and performance systems	Grammenos and Russell (1997), City of New York (1999), Russell and Moffat (2001), Langston and Shen (2007), Levine et al. (2007), Prowler (2008), Tobias and Vavaroutsos (2009), Caroon (2010), Gilder (2010), GBCA (2010)
Solar access: measures for summer and winter sun	Park (1998), City of New York (1999), Douglas (2006), Shaw et al. (2007), Dittmark (2008), Wilson and Ward (2009), GBCA (2010), Appleby (2011)

Table 12.1 (*Cont'd*).

Criterion	Previous research
Sense of place (social)	
Image/identity: social and cultural attributes, values, etc.	Wittkower (1988), ICOMOS (1994), Marquis-Kyle and Walker (1994), Curry (1995), Jokilehto (1996), Australian Government, Department of the Environment and Heritage (2004), Fournier and Zimnicki (2004), Harmon et al. (2006), Rodwell (2007), UNESCO (2007, 2009), NSW Department of Planning and RAIA (2008), Orbasli (2008), Bond and Charlemagne (2009), Yung and Chan (2012)
Aesthetics: architectural beauty, good appearance, proportion, etc.	ICOMOS (1994), Carter and Fortune (2008), Prowler (2008), Bond and Charlemagne (2009), Farrell (2010), GBCA (2010), Yung and Chan (2012)
Landscape/townscape: visual coherence and organisation of the built environment	Fournier and Zimnicki (2004), Zushi (2005), Davison, et al. (2006), Shaw et al. (2007), NSW Department of Planning and RAIA (2008)
History/authenticity: original fabric, timelessness, sociocultural traditions, practices, historic character or fabric, etc.	ICOMOS (1994), Marquis-Kyle and Walker (1994), Curry (1995), Jokilehto (1996), Australian Government, Department of the Environment and Heritage (2004), Fournier and Zimnicki (2004), Harmon et al. (2006), UNESCO (2007,2009), NSW Department of Planning and RAIA (2008), Orbasli (2008), Prowler (2008), Bond and Charlemagne (2009), Wilkinson et al. 2009, Yung and Chan (2012)
Amenity: provides comfort and convenience facilities	Graham (2005), Zushi (2005), Browne (2006), Fealy (2006), Peiser and Schmitz (2007), Prowler (2008)
Human scale: anthropometrics and fit to average human scale	Campbell (1996), Grammenos and Russell (1997), Russell and Moffat (2001)
Neighbourhood: local and social communities	HMSO (1987), Australian Government, Department of the Environment and Heritage (2004), Browne (2006), Carter and Fortune (2008), Prowler (2008), Yung and Chan (2012)
Quality standard (legal)	
Standard of finish: provision for high-standard workmanship	Osbourne (1985), Park (1998), Holborrow (2008), Whimster (2008)
Fire protection: provisions for fire safety	City of New York (1999), Queensland Government (2000), Davison et al. (2006), Douglas (2006), NSW Department of Planning and RAIA (2008), Solomon (2008), Horvath (2010)
Indoor environmental quality: provisions for nonhazardous materials, natural fabrics, etc.	City of New York (1999), Graham (2005), Prowler (2008), Rabun and Kelso (2009), Tobias and Vavaroutsos (2009), Caroon (2010), GBCA (2010)
Occupational health and safety: special needs of occupants, health and safety risks, building hazard and risk management plan	City of New York (1999), Queensland Government (2000), Douglas (2006), Levine et al. (2007), Carter and Fortune (2008), NSW Department of Planning and RAIA (2008), Prowler (2008), Caroon (2010), GBCA (2010), Horvath (2010)
Security: provision of direct and passive surveillance designs	Osbourne (1985), Douglas (2006), Carter and Fortune (2008), NSW Department of Planning and RAIA (2008), Prowler (2008), Solomon (2008)

(continued)

Table 12.1 *(Cont'd).*

Criterion	Previous research
Comfort: hygiene and clean environment, etc.	Osbourne (1985), Levine et al. (2007), Prowler (2008), Gilder (2010)
Disability access: provision for disability easement, facilities, etc.	Queensland Government (2000), Douglas (2006), NSW Department of Planning and RAIA (2008), Prowler (2008)
Energy rating: environmental performance measures	Douglas (2006), NSW Department of Planning and RAIA (2008), Atkinson et al. (2009), Reed et al. (2009), Schultmann et al. (2009), Tobias and Vavaroutsos (2009), GBCA (2010), Yudelson (2010), Appleby (2011)
Acoustics: noise control, sound insulation, etc.	Osbourne (1985), City of New York (1999), Davison et al. (2006), Douglas (2006), Levine et al. (2007), Wilkinson et al. 2009, Caroon (2010), Appleby (2011)
Context (political)	
Adjacent buildings: adjacent enclosures, vertical and visual obstacles	Davison et al. (2006)
Ecological footprint: appropriate measure of human carrying capacity	Balaras et al. (2004), Cantell (2005), Giles (2005), Langston and Shen (2007), UNEP (2007), Prowler (2008), Tobias and Vavaroutsos (2009), Gilder (2010)
Conservation: principles, guidelines, charters governing tangible and intangible heritage protection	ICOMOS (1994), Marquis-Kyle and Walker (1994), Curry (1995), Jokilehto (1996), Fournier and Zimnicki (2004), Harmon et al. (2006), UNESCO (2007, 2009), Prowler (2008), Yung and Chan (2012)
Community interest/participation: stakeholder relationship and support	HMSO (1987), Browne (2006), Peiser and Schmitz (2007), Langston et al. (2008), Prowler (2008)
Urban master plan: integrated skyline, urban landscape, built environment design and management/practice	Heath (2001), Douglas (2006), Peiser and Schmitz (2007), Wilson and Ward (2009)
Zoning: land uses and land patterns	Campbell (1996), City of New York (1999), Browne (2006), Douglas (2006), Peiser and Schmitz (2007), Wilkinson et al. 2009, Wilson and Ward (2009)
Ownership: collaborative commitment, sense of community or ownership, etc.	HMSO (1987), Peiser and Schmitz (2007), Whimster (2008)

computed as the standard deviation of actual responses divided by the standard deviation, assuming all responses were the same, so that no agreement = 0% and full agreement = 100%. This method of interpreting consensus was developed specifically for this research.

From this work it has been found that the seven obsolescence categories have reasonably equal weight – which was an assumption in the ARP model (Langston 2008) that now is vindicated. The coefficient of variation (CoV) of the seven criteria weights was just 8.32%. A scoring template has now been developed to enable new building design to be rated for future adaptation. This is illustrated in Figure 12.2.

Table 12.2 Weighting of obsolescence categories and assessment criteria.

AdaptSTAR criteria	Criterion weight	Consensus (%)	Total weight
Physical			16.08
Structural integrity and foundation	5.58	57.62	
Material durability and workmanship	5.33	60.66	
Maintainability	5.17	80.75	
Economic			13.40
Density and proximity	4.47	45.61	
Transport and accessibility	4.52	43.36	
Plot size and site plan	4.41	40.50	
Functional			15.23
Flexibility and convertibility	3.42	52.15	
Disassembly	2.96	43.36	
Spatial flow and atria	3.00	54.04	
Structural grid	3.03	48.58	
Service ducts and corridors	2.82	37.42	
Technological			14.85
Orientation and solar access	2.80	60.33	
Glazing and shading	2.54	44.27	
Insulation and acoustics	2.49	43.36	
Natural lighting and ventilation	2.67	48.17	
Energy rating	2.31	38.47	
Feedback on building performance and usage	2.04	40.00	
Social			14.37
Image and history	4.69	57.62	
Aesthetics and townscape	5.04	55.50	
Neighbourhood and amenity	4.64	50.60	
Legal			13.28
Standard of finish	4.36	44.72	
Fire protection and disability access	4.65	51.77	
Occupational health, IEQ, safety and security	4.27	45.17	
Political			12.79
Ecological footprint and conservation	4.05	39.50	
Community support and ownership	4.35	44.72	
Urban master plan and zoning	4.39	56.92	
		Total	100.00
		CoV	8.32%

12.4   International Case Studies

In order to illustrate the application of *adaptSTAR* in practice, the template shown in Figure 12.1 was used on ten well-documented adaptation projects sourced from around the world. The template was completed based on

Project Name

When nominating your opinion to EACH of the following statements, please assume that the latest adaptive reuse intervention has yet to occur. Your responses therefore relate to the latent conditions BEFORE such intervention. Please rate ALL statements using ONE opinion option and provide the key supporting REASON.

How do you judge the following statements for the above building/facility?	Strongly disagree	Disagree	Neutral	Agree	Strongly agree	What is the key reason that influenced your opinion?	Valid response ?
The building's foundations and frame have capacity for additional structural loads and potential vertical expansion.							X
The building fabric is well constructed using durable materials, providing potential retention of existing exterior and interior finishes.							X
The building currently has a low maintenance profile with modest expected levels of component repair and replacement over its remaining lifespan.							X
The building is situated in a bustling metropolis comprising mixed use development and proximity to potential markets.							X
The building is located near transport facilities and provides convenience for vehicular and pedestrian mobility.							X
The building enjoys a site with favourable plot size, access, topography, area, aspect and surrounding views.							X
The building's interior layout exhibits strong versatility for future alternative arrangements without significant disruption or conversion cost.							X
The building has significant components or systems that support disassembly and subsequent relocation or reuse.							X
The building has sufficient internal open space and/or atria that provides opportunity for spatial and structural transformations to be introduced.							X
The building has large floor plates and floor-to-floor heights with minimal interruptions from the supporting structure.							X
The building provides easy access to concealed ducts, service corridors and plant room space to ensure effective horizontal and vertical circulation of services.							X
The building is designed in such a way that it maximizes its orientation with good potential for passive solar strategies.							X
The building has appropriate fenestration and sun shading devices consistent with good thermal performance.							X
The building has an insulated external envelope capable of ensuring good thermal and acoustic performance for interior spaces.							X
The building is designed in ways that maximize daylight use and natural ventilation without significant mechanical intervention.							X
The building has low energy demand and is operating at or readily capable of achieving a 5-star Green Star® energy rating or equivalent.							X
The building supports efficient operational and maintenance practices including effective building management and control systems.							X
The building has developed strong intrinsic heritage values, cultural connections or positive public image over its life.							X
The building has high architectural merit including pleasing aesthetics and compatability with its surrounding streetscape.							X
The building provides relevant amenities and facilities within its neighbourhood that can add value to the local community.							X
The building displays a high standard of construction and finish consistent with current market expectations.							X
The building complies with current standards for fire prevention and safety, emergency egress and disability provisions.							X
The building offers an enhanced workplace environment that provides appropriate user comfort, indoor air quality and environmental health and safety.							X
The building's design is compatible with ecological sustainability objectives and helps minimize ongoing habitat disturbance.							X
The building displays a high level of community interest and political support for its future care and preservation.							X
The building's current or proposed future use conforms to existing masterplan, zoning and related urban planning specifications.							X

BUHREC Protocol Number RO-1208   **Score:**   **0.00**

Figure 12.2   Scoring template for *adaptSTAR* model.

latent conditions before the building conversion took place. The selected case studies comprised:

- 1881 Heritage, Hong Kong SAR (PRC)
- Peranakan Museum, City Hall (Singapore)
- Corso Karlín, Prague (Czech Republic)
- Arsenal de Metz, Metz (France)
- The Candy Factory Lofts, Toronto (Canada)
- Punta Della Dogana Contemporary Art Centre, Venice (Italy)
- Andel's Hotel, Lódz (Poland)
- Sugar Warehouse Loft, Amsterdam (The Netherlands)
- The Powerhouse, Long Island City (USA)
- John Knox Church, Melbourne (Australia)

The characteristics of each project are summarised in Sections 12.4.1.1–12.4.1.10. The selection of case studies is effectively random, sourced from a cross section of countries, possessing good levels of information available on the Internet. As all case studies are adaptive reuse conversions, this creates the possibility of comparing the outcome of *adaptSTAR* with Langston's ARP model (see Chapter 9). It is hypothesised that higher *adaptSTAR* scores lead to higher ARP scores in later life.

12.4.1   1881 Heritage, Hong Kong SAR (PRC)

Located in the heart of Tsim Sha Tsui (Hong Kong), the former Marine Police Headquarters has been rejuvenated and reintegrated into the urban fabric of the surrounding area. The 120+-year-old site has been carefully revitalised and transformed into a cultural and shopping landmark in Hong Kong. The renovation and conservation works were undertaken by Cheung Kong (Holdings) Limited. The site was home to the Hong Kong Marine Police from the 1880s until 1996, except for the period during World War II. The site comprises the Main Building, Stable Block, Time Ball Tower, Old Kowloon Fire Station and Fire Station Accommodation Block. The rich colonial buildings reflecting Victorian architecture were declared monuments under the Antiquities and Monuments Ordinance in 1994. The conversion, comprising a mix of old and new design, has won a host of industry awards and is arguably the most successful adaptive reuse project ever undertaken in Hong Kong. Further information can be found at http://www.1881heritage.com.

12.4.2   Peranakan Museum, City Hall (Singapore)

The architectural plans of Tao Nan School at 39 Armenian Street were drawn up and approved by the Municipal Engineer's Office in 1910. Construction of the building itself was completed in March 1912. The Tao Nan building was designed in the 'eclectic classical' style. The fluted columns and the symmetry of the building are characteristic of classical architecture, while the balconies fronting the façade suggest a colonial or tropical style. The layout of the building is also based on Straits Settlements bungalows with rooms arranged around a common central hall and toilets and kitchens outside the principal building. In 1976, it was decided that Tao Nan should move from the city to cater to the rising number of pupils in the suburbs, where a new school building could also provide better modern facilities, and in 1982 Tao Nan surrendered its Armenian Street premises and relocated to Marine Parade. Appropriately, for a building that was once a Chinese school, the permanent exhibition of the Asian Civilisations Museum (ACM) began with a focus on different aspects of Chinese culture and civilisation, ranging from architecture to the connoisseurship of the literati. ACM closed at the end of 2005 to be redeveloped as a new museum to showcase the eclectic Peranakan culture. Today, the old Tao Nan School has entered the latest and most colourful phase in its history – as the Peranakan Museum. Further information can be found at http://www.peranakanmuseum.sg/themuseum/historyofbuilding.html.

12.4.3   Corso Karlín, Prague (Czech Republic)

As occurs in many European cities, former industrial buildings have lost their original purpose and must be transformed to accommodate other uses. This is the case of Corso Karlín, an office block that forms part of an ambitious development plan for the entire Prague 8 district. The intent was to modernise a quarter without depriving it of its historical roots: Corso Karlín is an example of a former industrial building that has been transformed into a modern, efficient commercial work centre of about 7000 m². As part of the Real Estate Karlín Group's plan to modernise the district, the architectural practice of Taller de Arquitectura used its previous adaptive reuse experience to complete the renovation of Corso Karlín in 2001. The original building has been preserved and its base renovated. The new roof is of glass, and red stucco arcades highlight the composition of the existing ground floor. The aim here was to maintain a dialogue between light and shadow, between solidity and transparency and between the language of classical architecture and modern materials. The area, once marked by dirty industrial spaces, is now breathing easier with a new life focused on business. The conversion of Corso Karlín plays out this conversion, opening up the building to its surrounding area and bringing in lots of natural light. By reusing the building, the designers have been able to solidify the district's past as well as future. Further information can be found at http://www.ricardobofill.com/EN/630/PROJECTS/Corso-I.html.

12.4.4   Arsenal de Metz, Metz (France)

Built in 1863 during the reign of Napoleon III, this building served as a military arsenal for over a century. The restoration of the building, comprising about 10000 m² of built surface, was directed at accommodating a rehearsal hall; a concert hall for chamber music; a restaurant; exhibition gallery; offices for administration, management and centre services; and a 1500-seat auditorium. One wing of the building, originally square with a 30 x 50-m internal courtyard, has been sacrificed in order to open up the central courtyard to the city, forming a public square and giving a better view of the Knight Templars Chapel, which dates back to the twelfth century. The façade has been slightly modified by means of cladding with slabs of natural stone with metal joints that underline the rhythm of the arches. The introduction of big new windows has lightened the heavy, opaque solidity of the old military building. The main auditorium is underground, situated beneath the central square. The roof, with its wooden structure covered with anodised steel, is flat – the problems of reverberation were resolved by means of a design based on detailed studies of acoustic performance. The hall has two ramped seating areas; the smaller, with a pronounced incline, can be used to accommodate the choir when necessary. The orchestra pit is located between these two seating areas, on the lowest level of the auditorium. With its conversion completed in 1989, the building is now home to the Symphony Orchestra of Lorraine. This project has helped to open up the space to the public and provided a new cultural venue, built upon its storied past. Further information can be found at http://www.architizer.com/projects/arsenal-music-center/.

12.4.5   The Candy Factory Lofts, Toronto (Canada)

Dating back to the 1930s, this warehouse is located west of downtown Toronto in the West Queen Street neighbourhood. It previously had been used as a Ce De Candy Company factory, the makers of Smarties. In 1999, it was adaptively reused as residential lofts. Quadrangle Architects and the Metro Ontario Group created new loft residences out of the disused factory, which is arguably Toronto's most famous hard loft building, so the building now provides a great housing option in a neighbourhood of restaurants, art galleries and bars. The six-storey post and beam loft building now houses 121 individual loft units. Some of the amenities enjoyed by the residents include 24-h concierge service, a party room, a guest suite, fitness room, roof terraces and wide hallways. Further information can be found at http://www.mrloft.ca/Loft-Buildings/Candy-Factory-Lofts-993-Queen-St-W-Toronto.

12.4.6   Punta Della Dogana Contemporary Art Centre, Venice (Italy)

For centuries, this rusticated-stone and plaster-on-brick building served as the Customs House in Venice, Italy. The seventeenth-century building, located at the eastern tip of Dorsoduro Island and next to Longhena's domed basilica of Santa Maria della Salute, was shut down and left vacant in the 1970s. French billionaire and art collector François Pinault won the bid to convert the building into a contemporary art museum. He owns (through his foundation) Palazzo Grassi in Venice and has one of the world's largest collections of contemporary art (nearly 2500 pieces).

Japanese architect Tadao Ando was selected by Pinault to bring the conversion to life. Known for his creative use of natural light and for architecture that follows the natural forms of the landscape, Ando's approach to architecture was once categorised as critical regionalism. He has focused his work in Japan but has a number of projects in Europe as well as the US. The adaptive reuse project took 14 months to complete and has created a lasting impression on this significant site in Venice. While the building itself is triangular and matching the shape of the island, the interior has been divided up into long rectangles for a number of different galleries. The façade was completely restored and all openings were replaced. A protective shell at the building's base was installed to secure it against high water up to about 2 m, and the brick foundation was replaced. Skylights were installed, while the wooden roof trusses were recovered and the roof itself fully restored. The $28 million project was opened on 6 June 2009. Further information can be found at http://www.designboom.com/architecture/tadao-ando-punta-della-dogana-museum-in-venice/.

12.4.7   Andel's Hotel, Lódz (Poland)

The weaving mill, built in 1852 by textile magnate Izrael Poznanski, is defined by its red-brick exterior and cast-iron pillars. After over a century of use, the complex was abandoned in the 1990s. This adaptive reuse project, completed in 2009, was commissioned by Warimpex Finanz-und Beteiligungs AG of Vienna. The interior design was handled by Jestico + Whiles, which is famous for their practical, innovative and contemporary solutions. The executive architect for the project was OP Architekten, founded by the architects Orlinski and Poplawski. The former factory has been transformed into the first 4-star hotel in Lódz, Poland. Jestico + Whiles painstakingly followed the city's strict codes of historic building preservation to honour the tradition of the building. The hotel is actually one piece to the larger reuse of the complex, now called Manufaktura (a retail and entertainment centre). The 20000-m^2 four-level hotel includes 180 guestrooms and 80 long-stay apartments. The hotel lobby is marked by the building's original cast-iron pillars supporting the red-brick vaulted roof and three light wells that slice through the ceiling with sculptural displays of concentric circles denoting the balustrades of each floor above, each lit with changing coloured LEDs. The hotel's pool was created out of a nineteenth-century fire water storage tank and is located in a cantilevered glass box on the top floor, overhanging the building's brick exterior. Further information can be found at http://www.vi-hotels.com/en/andels-lodz/.

12.4.8   Sugar Warehouse Loft, Amsterdam (The Netherlands)

The old sugar warehouse was built in 1763. It is located on the Bloemgracht (Flower Canal) in Amsterdam. The building and its interior have largely gone untouched/unmodified for about 250 years. George Gottl and Oliver

Michell are both the owners and residents of the loft as well as the owners and directors of UXUS Design, which handled the reuse of the interiors. In keeping the original open-plan layout, they created a series of dramatic curtain walls that could be opened and closed according to the needs of use. Made of luxurious Italian linen, the curtain walls are opaque when lit from the front yet transparent from behind. At night, the space becomes a series of glowing tents, creating the effect of a surreal interior landscape. Located on one of the ever-popular canals, this loft conversion offers its residents great access and views of the city. The conversion was completed in 2003, and the building has a new life as a private residence. Further information can be found at http://loftlifemag.com/mu/?p=3890.

12.4.9   The Powerhouse, Long Island City (USA)

The Long Island power station was built in 1906 to further enhance New York City's transportation network. It served to power and expand the Long Island and Pennsylvania Railroads. Georgia O'Keefe used the building as a centrepiece to her painting 'Across the East River' in the 1920s. The building was vacant for a decade and used as a plumbing warehouse. The vacant structure was purchased in 2004 by CGS Developers. The year prior, they had passed up buying the site but changed course when the city rezoned the area for residential development. Along with the steam plant, the group also bought the neighbouring Schwartz Chemical Plant to be included in the redevelopment. CGS successfully converted the old plant into upscale residential condominiums in 2008. The adaptive reuse of the old plant is just the first step in the project's three phases. This first phase provides 177 living units. The four steam stacks were removed due to structural issues, but glass towers were put in their place and provide additional living space. CGS's intent to reuse the building, however, seems to have conflicting reports. Initially, the firm had planned to demolish the entire building and put four separate buildings in its place. One report claims reuse of the building was necessary due to the high costs of demolition, while another report states demolition would have saved money but that the public outcry against destruction made CGS change their mind. In the end, however, the building was converted and extended. The desire is that this redevelopment and repurposing will help to spur additional investment in the community. Further information can be found at http://thepowerhouselic.com/.

12.4.10   John Knox Church, Melbourne (Australia)

The John Knox Church demonstrates one way that we as a society can maximise the sustainable use of existing resources by repurposing the building stock that we have inherited from previous users. The adaptive reuse of our existing buildings also maintains an important connection with familiar things, enriching memory and place in a community. Williams Boag Architects has not only conserved the early building fabric of this heritage-listed National Trust church; they transformed its essence away from its nominal purpose as a religious building focused on worship. In 2009 it was converted into a contemporary family dwelling, with its polyphonic, ambiguous mix of function, utility, privacy and amenity. This 1867 historic building and its interior rich in its detailed and striking stained glass are offered a new social and physical context in which it sits comfortable. Further information can be found at http://www.australiandesignreview.com/architecture/1950-knox-church-residence.

12.5   Discussion

Of course, *adaptSTAR* did not exist at the time any of the case studies were conceived, so for the purposes of this research, the scores are derived with the benefit of hindsight. Nevertheless, the expected conditions at the time of design are used to inform the assessment. While it is reasonable to hypothesise a positive correlation between the *adaptSTAR* score (determined using the template in Figure 12.1) and the ARP score (determined using the template in Figure 9.4) for each of the previous case studies, it is to some extent a self-fulfilling prophecy. This is because the very reason the adaptive reuse conversions were undertaken is because the latent conditions were judged to be conducive to a successful outcome. If this were not the case, then the projects would surely have encountered a different fate.

Based largely on preliminary data available in the public domain, computed values are provided in Table 12.3 and their correlation shown in Figure 12.3. More detailed on-site investigations are planned for the near future to confirm and validate the data.

The *adaptSTAR* scores were found to be reasonably high, ranging from 55.2 (John Knox Church) to 85.6 (Arsenal de Metz), with a mean of 70.3 and a CoV of just 15%. This suggests what is already understood – each

Table 12.3   Summary of international case study evaluation.

Case study	Year constructed	Year converted	AdaptSTAR score (%)	ARP score (%)
1881 Heritage, Hong Kong SAR (PRC)	1884	2009	74.9	59.2
Peranakan Museum, City Hall (Singapore)	1912	2005	73.5	57.6
Corso Karlín, Prague (Czech Republic)	1891	2001	83.8	60.2
Arsenal de Metz, Metz (France)	1863	1989	85.6	74.3
The Candy Factory Lofts, Toronto (Canada)	1930	1999	75.9	58.4
Punta Della Dogana Contemporary Art Centre, Venice (Italy)	circa 1600	2009	69.3	28.7
Andel's Hotel, Lódz (Poland)	1852	2009	68.3	59.8
Sugar Warehouse Loft, Amsterdam (The Netherlands)	1763	2003	55.8	33.4
The Powerhouse, Long Island City (USA)	1906	2008	60.4	52.4
John Knox Church, Melbourne (Australia)	1867	2009	55.2	59.1
		Mean	*70.3*	*54.3*
		CoV	*15%*	*25%*

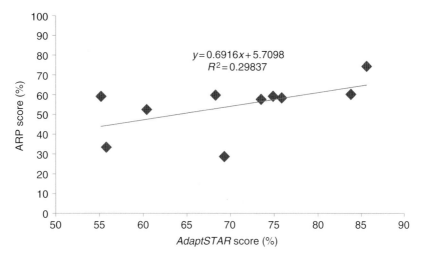

Figure 12.3   Correlation test between *adaptSTAR* and ARP models.

case study had an original design with a natural propensity for future adaptation. The year in which the adaptation took place is the year used in the ARP calculation. Overall they had a wider range and dispersion, with two projects in particular scoring quite low due to their age. The motivation for these two adaptations is arguably more to do with ensuring their survival than an opportunity for redevelopment.

A weak relationship ($r^2 = 0.30$) between *adaptSTAR* and ARP scores is evident, but the trend line supports the hypothesis that both are positively correlated. Using linear regression (line of best fit), an equation is produced that defines the relationship. ARP scores (y) can be predicted from *adaptSTAR* scores (x) and vice versa. However, with only ten data points, this result is preliminary, and further testing is warranted. Note also that Punta Della Dogana Contemporary Art Centre is a potential outlier as it exceeds the age range of the ARP model (i.e. 300 years) and demands a manually assessed physical life (in this case 500 years was chosen), which was determined on the basis that it had survived for more than 400 years already. If this project is removed from the analysis, the resultant r^2 value increases to 0.50.

However, it is arguably unfair to compare *adaptSTAR* and ARP on this basis. Projects that are converted prematurely or belatedly in the life cycle will have a diminished ARP that understates their true potential. So it is more appropriate to assume the conversion takes place at the optimum time (i.e. when building age and useful life are equal). When this assumption is enacted, and without the need to remove any outliers, the r^2 value rises to 0.70, which is considered indicative of a strong relationship. This is illustrated in Figure 12.4.

But not all adaptations are successful when judged by a range of stakeholders. In some cases the investment in their conversion is made on the basis of love rather than return. In other cases, adaptation is the lesser of two evils, given that heritage controls may prohibit demolition and impending building dereliction introduces financial liabilities and increased risk to

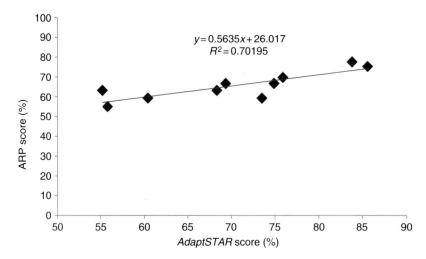

Figure 12.4   Adjusted correlation test between *adaptSTAR* and ARP models.

the owner, putting downward pressure on property values (Conejos et al. 2013). However, a big advantage of such projects is usually their embodied energy saving due to substantial reuse of materials in place.

12.6   Conclusion

It is appropriate to consider future 'adaptivity' of buildings during the design process given our current knowledge that we have access to a finite amount of natural resources and we can no longer construct buildings as if they were disposable consumer goods. The philosophy of 'long life, loose fit and low energy' taught for centuries in architectural schools can be translated as 'durable, adaptable and sustainable' in modern parlance. Success in this endeavour demands tools that can assist in identifying appropriate design and guiding better decisions about proposed interventions to ensure that the contributions that built assets make to society are maximised.

References

Abbotts, J., Ertell, K.B., Leschine, T.M. and Takaro, T.K. (2003) Building leasing at the department of energy's Hanford site: lessons learned from commercial reuse, *Federal Facilities Environmental Journal*, 14 (Spring), 95–107.

Anonymous (2006) Sustainable solar solutions case study 02, Sustainability Victoria, Melbourne, 3. Available at http://www.sustainability.vic.gov.au/www/html/1589-case-studies.asp. Accessed on 10 June 06.

Appleby, P. (2011) *Integrated sustainable design of buildings*, Sterling, VA: Earthscan.

Arge, K. (2005) Adaptable office buildings: theory and practice, *Facilities*, 23(3–4), 119–127.

Atkinson, B. (1988) Urban ideals and the mechanism of renewal, in proceedings of RAIA Conference, Sydney, Australia, June 88.

Atkinson, C., Yates, A. and Wyatt, M. (2009) *Sustainability in the built environment: an introduction to its definition and measurement*, UK: BRE Press.

Australian Government, Department of the Environment and Heritage. (2004) *Adaptive reuse: preserving our past, building our future*, Canberra: Commonwealth of Australia, Department of Environment and Heritage. Available at http://www.environment.gov.au/heritage/publications/protecting/pubs/adaptive-reuse.pdf. Accessed on 21 August 2013.

Balaras, C.A., Dascalaki, E. and Kontoyiannidis, S. (2004) Decision support software for sustainable building refurbishment, *ASHRAE Transactions*, 110(1), 592–601.

Ball, R. (1999) Developers, regeneration and sustainability issues in the reuse of vacant industrial buildings, *Building Research and Information*, 27(3), 140–148.

Bond, S. and Charlemagne, D.W. (2009) Built cultural heritage and sustainability: the role of value based decisions, in proceedings of EU Sustainable Energy Week, February 09.

Brandt, M. (2006) How to adaptively reuse a community asset? *Heritage: the Magazine of the Heritage Canada Foundation*, 9(2), 21–22.

Browne, L.A. (2006) Regenerate: reusing a landmark building to economically bolster urban revitalization, Master's Thesis, University of Cincinnati, OH.

Campbell, J. (1996) Is your building a candidate for adaptive reuse? *Journal of Property Management*, 61(1), 26–29.

Cantell, S.F. (2005), The adaptive reuse of historic industrial buildings: regulation barriers, best practices and case studies, Master's Thesis, Virginia Polytechnic Institute and State University, Blacksburg, VA.

Caroon, J. (2010) *Sustainable preservation: greening existing buildings*, Hoboken, NJ: John Wiley & Sons.

Carter, K. and Fortune, C. (2008) *A consensual sustainability model: a decision support tool for use in sustainable building project procurement*, London: RICS Research.

Chusid, M. (1993) Once is never enough, *Building Renovation*, (March–April), 7–20.

City of New York. (1999) *High performance building guidelines*, New York: NYCDDC.

Conejos, S. (2013) Designing for future building adaptive reuse, PhD Thesis, Bond University, Gold Coast, Australia.

Conejos, S., Langston, C. and Smith, J. (2013) AdaptSTAR model: a climate-friendly strategy to promote built environment sustainability, *Habitat International*, 37(1), 95–103.

Curry, M. (1995) Archaeological resource management in the UK: an introduction, Book Review, *International Journal of Cultural Property*, 4(1), 179–181.

Davison, N., Gibb, A.G., Austin, S.A., Goodier, C.I. and Wagner, P. (2006) The multispace adaptable building concept and its extension into mass customisation, in proceedings of Adaptables2006 International Conference on Adaptable Building Structures, Eindhoven, The Netherlands, 3–5 July 2006.

Dittmark, H. (2008) Continuity and context in urbanism and architecture: honesty of a living tradition, *Conservation Bulletin*, 59 (Autumn), 7–9.

Doak, J. (1999) Planning for the reuse of redundant defense estate: disposal processes, policy frameworks and development impacts, *Planning Practice and Research*, 14(2), 211–224.

Douglas, J. (2006) *Building adaptation* (2nd edition), London: Elsevier.

DSE. (2008), *PRISM*, Melbourne: Department of Sustainability and Environment.

Farrell, A. (2010) Intelligent eco-physiological architecture: a primer for a sustainable built environment, in proceedings of Building a Better World: CIB World Congress 2010, Salford, UK, 10–13 May 2010.

Fealy, J. (2006) Adaptive reuse for multi-use facilities in an urban context: making the city home again, Master's Thesis, University of Cincinnati, OH.

Fournier, D. and Zimnicki, K. (2004) *Integrating sustainable design principles into the adaptive reuse of historical properties*, Washington, DC: US Army Corps of Engineers.

Gallent, N., Howet, J. and Bellt, P. (2000) New uses for England's old airfields, *Area*, 32(4), 383–394.

GBCA. (2010) What is green star? Available at http://www.gbca.org.au/green-star/green-star-overview/. Accessed on 11 August 2013.

Gilder, J. (2010) Bio inspired intelligent design for the future of buildings, in proceedings of Building a Better World: CIB World Congress 2010, Salford, UK, 10–13 May 2010.

Giles, G. (2005) Adaptive reuse in an urban setting: evaluating the benefits of reusing an existing building site in Florida for maximum profit potential and eco-effectiveness, *Environmental Design and Construction*, 8(3), 72.

Gorse, C. and Highfield, D. (2009) *Refurbishment and upgrading of buildings* (2nd edition), New York: Spon Press.

Graham, P. (2005) Design for adaptability, *BDP Environment Design Guide*, 66, 1–9.

Grammenos, F. and Russell, P. (1997) Building adaptability: a view from the future, in proceedings of the Second International Conference on Buildings and the Environment, Paris, France, June 97.

Habraken, N. (1998) *The structure of the ordinary: form and control in the built environment*, Cambridge, MA: MIT Press.

Harmon, D., Mcmanamon, F.P. and Pitcaithley, D.T. (2006) *The Antiquities Act: a century of American archaeology, historic preservation and nature*, Tucson, AZ: University of Arizona Press.

Heath, T. (2001) Adaptive re-use of offices for residential use: the experiences of London and Toronto, *Cities*, 18(3), 173–184.

HMSO. (1987) *Town and country planning (use classes)*, London: HMSO.

Holborrow, W. (2008) Cutting down on carbon from the public sector estate, *Conservation Bulletin*, 57 (Spring), 26–29.

Horvath, R.J. (1994) National development paths 1965–1987: measuring a metaphor, *Environment and Planning A*, 26, 285–305.

Horvath, T. (2010) Necessity of modernization of modern buildings, in proceedings of Building a Better World: CIB World Congress 2010, Salford, UK, 10–13 May 13.

ICOMOS. (1994) The Nara Document on authenticity. Available at http://www.icomos.org. Accessed on 8 August 2013.

Jacobs, J. (1961) *The death and life of great American cities*, New York: Random House.

Jokilehto, J. (1996) A history of architectural conservation: the contribution of English, French, German and Italian towards an international approach to the conservation of cultural property, PhD Thesis, University of York, UK.

Knaack, U. and Klein, T. eds. (2009) *The future envelope 2: architecture, climate, skin*, Amsterdam: IOS Press.

Langston, C. (2008) The sustainability implications of building adaptive reuse (keynote paper), in proceedings of CRIOCM2008, Beijing, China, October 31–November 1, 00, pp. 1–10.

Langston, C. and Shen, L.Y. (2007) Application of the adaptive reuse potential model in Hong Kong: a case study of Lui Seng Chun, *The International Journal of Strategic Property Management*, 11(4), 193–207.

Langston, C., Wong, F., Hui, E. and Shen L.Y. (2008) Strategic assessment of building adaptive reuse opportunities in Hong Kong, *Building and Environment*, 43(10), 1709–1718.

Lehmann, S. (2010) Low- to-no carbon city: is there a lesson from Potsdamer Platz Berlin for the Rapid Urban Transformation of Shanghai?, in proceedings of International Conference on Sustainable Urbanization: ICSU 2010, Hong Kong, China, 12–17 December 2010.

Lenzen, M. and Treloar, G.J. (2003) Differential convergence of life-cycle inventories towards upstream production layers: implications for life-cycle assessment, *Journal of Industrial Ecology*, 6(3–4), 137–160.

Leone, A. (2003) Late antique North Africa: production and changing use of buildings in urban areas, *Al-Masáq*, 15(1), 21–33.

Levine, M., Urge-Vorsatz, D., Blok, K., Geng, L., Harvey, D., Land, S., Levermore, G., Mongameli Mehlwana, A., Mirasgedis, S., Novikova, A., Rlling, J. and Yoshino, H. (2007) Residential and Commercial Buildings, IPCC Fourth Assessment Report: Climate Change 2007, Cambridge: Cambridge University Press.

Maggs, A. (1999) Adaptive reuse, *Place*, 1(4), 33–34.

Marquis-Kyle, P. and Walker, M. (1994) *The illustrated Burra Charter: making good decisions about the care of important places Australia*, Sydney: ICOMOS with the assistance of the Australian Heritage Commission.

McLaren, P. (1996) *Adaptation and reuse, monuments and sites Australia: Australia ICOMOS*, Sri Lanka National Committee of ICOMOS, pp. 170–176.

Nakib, F. (2010) Toward an adaptable architecture guidelines to integrate adaptability in building', in proceedings of Building a Better World: CIB World Congress 2010, Salford, UK, 10–13 May 2010.

Ness, D. and Atkinson, B. (2001) Re-use/upgrading of existing building stock, in *BDP Environment Design Guide*, DES May 11, Canberra: Building Design Professions.

Newman, H.K. (2001) Historic preservation policy and regime politics in Atlanta, *Journal of Urban Affairs*, 23(1), 71–86.

NSW Department of Planning and RAIA (2008) *New uses for heritage places: guidelines for the adaptation of historic buildings and sites*, Parramatta: Heritage Council of New South Wales.

Orbasli, A. (2008) *Architectural conservation*, Malden, MA: Blackwell Science.

Osbourne, D. (1985) *Introduction to building*, Mitchell's Building Series, Batsford Academic and Educational, London: Longman Scientific & Technical.

Park, S.C. (1998) Sustainable design and historic preservation, *CRM*, 2, 13–16.

Peiser, R.B. and Schmitz, A. eds. (2007) *Regenerating older suburbs*, Washington, DC: Urban Land Institute.

Pevsner, N. (1975) *Pioneers of modern design: from William Morris to Walter Gropius*, Middlesex: Penguin Books.

Poon, B.H.S. (2001) Buildings recycled: city refurbished, *Journal of Architectural Education*, 54(3), 191–194.

Prowler, D. (2008) *Whole building design guide*, Washington, DC: National Institute of Building Sciences.

Queensland Government (2000) *Ecologically sustainable design in office fit out*, Australia: Queensland Department of Public Works.

Rabun, S. and Kelso, R. (2009) *Building evaluation for adaptive reuse and preservation*, Hoboken, NJ: John Wiley and Sons.

Reed, R., Bilos, A., Wilkinson, S. and Schulte, K.W. (2009) International comparison of sustainable rating tools, *Journal of Sustainable Real Estate*, 1(1), 1–22.

Remøy, H., Koppels, P.W. and de Jonge, H. (2009) Keeping up appearance, *Real Estate Research Quarterly*, 1(3), 25–30.

Rodwell, D. (2007) *Conservation and sustainability in historic cities*, Oxford Blackwell Publishing.

Russell, P. and Moffat, S. (2001) Adaptability of buildings, in proceedings of IEA Annex 31, November 2001, pp. 1–13.

Schultmann, F., Sunke, N. and Kruger, P.K. (2009) Global performance assessment of buildings: a critical discussion of its meaningfulness, in proceedings of SASBE 2009, Delft, Netherlands.

Shaw, R., Colley, M. and Connell, R. (2007) *Climate change adaptation by design: a guide for sustainable communities*, London: TCPA.

Siddiqi, K. (2006) Benchmarking adaptive reuse: a case study of Georgia, *Environmental Technology and Management*, 6(3–4), 346–361.

Solomon, R. (2008) Measuring optimum and code-plus design criteria for the high rise environment, in proceedings of CTBUH 8th World Congress, Dubai, March 08.

Tobias, L. and Vavaroutsos, G. (2009) *Retrofitting office buildings to be green and energy-efficient: optimizing building performance, tenant satisfaction and financial return*, Washington, DC: Urban Land Institute.

UNEP (2007) *Buildings and climate change status: challenges and opportunities*, Paris: UNEP publications.

UNEP (2009) *Buildings and climate change: summary for decision-makers*, Paris: UNEP Publications.

UNESCO (2007) *Asia conserved: lessons learned from the UNESCO Asia-Pacific Heritage Awards for Culture Heritage Conservation*, Bangkok: UNESCO.

UNESCO (2009) *Hoi An protocols for best conservation practice in Asia: professional guidelines for assuring and preserving the authenticity of heritage sites in the context of the cultures of Asia*, Bangkok: UNESCO.

Vakili-Ardebili, A. (2007) Complexity of value creation in sustainable building design (SBD), *Journal of Green Building*, 2(4), 171–181.

Van Driesche, J. and Lane, M. (2002) Conservation through conversation: collaborative planning for reuse of a former military property in Sauk County, Wisconsin, *Planning Theory and Practice*, 3(2), 133–153.

Whimster, R. (2008) Inventing the future, *Conservation Bulletin*, 57 (Spring), 20–25.

Wilkinson, S.J., James, K. and Reed, R. (2009) Using building adaptation to deliver sustainability in Australia, *Structural Survey*, 27(1), 46–61.

Wilson, A. and Ward, A. (2009) Design for adaptation: living in a climate changing world. Available at http://www.buildinggreen.com/auth/article.cfm/2009/8/28/Designfor-adaptation-living. Accessed on 15 Sept 2009.

Wittkower, R. (1988) *Architectural principles in the age of humanism* (4th edition) New York: St. Martin's Press.

Yudelson, J. (2010) *Greening existing buildings*, New York: McGraw Hill.

Yung, E.H.K. and Chan, E.H.W. (2012) Implementation challenges to the adaptive reuse of heritage buildings: towards the goals of sustainable, low carbon cities, *Habitat International*, 32, 352–361.

Zeiler, W., Quanjel, E., Velden, J. and Wortel, W. (2010) Flexible design process innovation: integral building design method, in proceedings of Building a Better World: CIB World Congress 2010, Salford, UK, 10–13 May 2010.

Zushi, K. (2005) Potential residential buildings for adaptive reuse: Cincinnati's CBD, Master's Thesis, University of Cincinnati, OH.

Index

Sustainable Building Adaptation: Innovations in Decision-Making, First Edition.
Sara J. Wilkinson, Hilde Remøy and Craig Langston.
© 2014 John Wiley & Sons, Ltd. Published 2014 by John Wiley & Sons, Ltd.

WILEY Blackwell

Also available from Wiley-Blackwell

Sustainable Refurbishment
Shah
9781405195089

Facilities Change Management
Finch
9781405153461

Facilities Manager's Desk
Reference, 2ed
Wiggins
78111846294

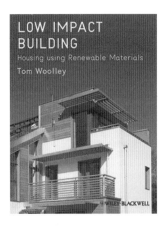

Low Impact Housing: Building
with Renewable Materials
Woolley
9781444336603

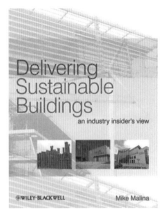

Delivering Sustainable Buildings
Malina
9781405194174

International Facility
Management
Roper & Borello
9780470674000

other books of interest

Enhancing Building Performance
Mallory-Hill, et al.
9780470657591

Legal Concepts for Facilities Managers
Thomas-Mobley
9780470674741

www.wiley.com/go/construction